TV COPS

The police drama has been one of the longest running and most popular genres in American television. In *TV Cops*, Jonathan Nichols-Pethick argues that, perhaps more than any other genre, the police series in all its manifestations–from *Hill Street Blues* to *Miami Vice* to *The Wire*–embodies the full range of the cultural dynamics of television.

Exploring the textual, industrial, and social contexts of police shows on American television, this book demonstrates how police dramas play a vital role in the way we understand and engage issues of social order that most of us otherwise experience only in such abstractions as laws and crime statistics. And given the current diffusion and popularity of the form, we might ask a number of questions that deserve serious critical attention: Under what circumstances have stories about the police proliferated in popular culture? What function do these stories serve for both the television industry and its audiences? Why have these stories become so commercially viable for the television industry in particular? How do stories about the police help us understand current social and political debates about crime, about the communities we live in, and about our identities as citizens?

Jonathan Nichols-Pethick is Associate Prefessor of Media Studies at DePauw University. His work has appeared in *The Velvet Light Trap*, *Cinema Journal*, and in the anthology *Beyond Prime Time: Television Programming in the Post-Network Era*.

TV COPS

The Contemporary American Television Police Drama

Jonathan Nichols-Pethick

NEW YORK AND LONDON

First published 2012
by Routledge
711 Third Avenue, New York, NY 10017

Simultaneously published in the UK
by Routledge
2 Park Square, Milton Park, Abingdon, Oxon OX14 4RN

Routledge is an imprint of the Taylor & Francis Group, an informa business

Library of Congress Cataloging in Publication Data
Nichols-Pethick, Jonathan.
TV cops : the contemporary American television police drama / Jonathan Nichols-Pethick.
p. cm.
Includes bibliographical references and index.
1. Detective and mystery television programs–United States–History and criticism. 2. Television cop shows–United States–History and criticism.
I. Title.
PN1992.8.D48N53 2012
791.45'6556—dc23 2011045116

ISBN: 978-0-415-87787-9 (hbk)
ISBN: 978-0-415-87788-6 (pbk)
ISBN: 978-0-203-85502-7 (ebk)

Typeset in Bembo
by Cenveo Publisher Services

For Nancy, David, and Trinity

In Memory of Kathryn Lasky

CONTENTS

ACKNOWLEDGMENTS

The process of writing this book has put me in mind of a scene from Nick Hornby's novel *High Fidelity* in which Rob Fleming's girlfriend praises him for producing something real instead of continuing in his role as a "professional appreciator." But like anyone who puts something out there, I haven't done this alone. The following people deserve my deepest thanks and, well, appreciation.

Matthew Byrnie first encouraged this project during his tenure at Routledge. Erica Wetter along with Margo Irvin took it on when Matt left. Sioned Jones guided the manuscript through the final production stages. Dan Harding's attention to detail and style made every page of this book better. Each of you has shown me a tremendous amount of patience and kindness over the years that it took me to get the manuscript completed and into your hands. I couldn't have worked with a smarter, more professional, insightful, and encouraging group of people.

Perhaps more than any other person, Chris Anderson has helped guide this project with patience, wisdom, and stamina. From his first enthusiastic encouragement of my impulse to write about cop shows (fifteen years ago), to the moment I completed the final draft of this book, Chris has been the central figure in my academic career, and a great friend. Michael Curtin has also been with this project from the beginning. His wealth of ideas continues to push my thinking about television and film in wonderful new directions. There are parts of this book that would literally have been impossible without him, not the least of which was the very last push. Over breakfast one day in Santa Barbara, Michael posed a "what if" question to me. The possibility he laid out clarified my project in ways I couldn't myself. From that moment on, I knew I could finish the book. Joan Hawkins has been one of the best friends and mentors I could hope for: enormously generous with her time, energy, encouragement, and insights. And one of the smartest and nicest people I've ever met. Period. Jim Naremore was also instrumental in helping me get the project off the ground and through its early stages. Jim's own work is a model for any and all academics to follow, and his ability to generate profoundly useful advice from the slightest slivers of my ideas has made me a much stronger writer and thinker.

Many friends and colleagues have lived through this with me, and offered me guidance and support along the way, whether they know it or not: Amanda Lotz, Tim Havens, Jon Kraszewski, Sue Kraszewski, Bjorn Ingvoldstadt, Lori Hitchcock, Sherra Schick, Darrell Enck-Wanzer, Suzanne Enck-Wanzer, Chris Dumas, Bob Rehak, Jim Kendrick, Tonia Edwards, Jeff Bennet, Claire Sisco-King, David Moscowitz, Kristy Horn-Sheeler, Jason Mittell, Jonathan Gray, Serra Tinic, Sharon Ross, Todd Avery, and Gigi Thibodeau. Each of you has helped me get to the finish line in one way or another, and I thank you.

My colleagues at DePauw University have been patient and generous. I would especially like to thank Jennifer Adams, Meryl Altman, Susan Anthony, Samuel Autman, Dave Bohmer, Joyce Christiansen, Ron Dye, Melanie Finney, Seth Friedman, Tim Good, Andrew Hayes, Tiffany Hebb, Kevin Howley, Geoff Klinger, Jeff McCall, Kent Menzel, Keith Nightenhelser, Kerry Pannell, Pam Propsom, Jackie Roberts, Bob Steele, Steve Timm, Sheryl Tremblay, Carrie Van Brunt, Chris White, Susan Wilson, and Dave Worthington. Each of you, in your own way, makes DePauw a great place to work.

Thank you as well to Julie Martin, Tom Fontana, David Chambers, and Julie Chambers for generously allowing me access to your wisdom and insights as writers and producers of some of the best television around.

I owe special gratitude to my family. My mother and father have supported me in every possible way in everything I've done. From baseball diamonds to drum sets to this book, they have been (and continue to be) the guiding force behind me, and my biggest supporters (and they know way more about police dramas than I do). Everyone should be so lucky to have parents like mine. I also want to thank my older brothers, Chris and Steve, who have always led the way and cleared the path for me. It gives me no end of pleasure to know that we are close friends as well as brothers. My in-laws, especially Dorothy Nichols, Susan Pethick, Pam Pearce, Bob and Amy Nichols, Cathy and Paul Miller, Jim and Jeanne Nichols, and John and Sue Austin, as well as my two nieces, Lindsay and Flora Pethick, also deserve thanks and recognition for being part of this process in one way or another.

And I want to thank Godiva (Miss G.) for all the companionship while I typed away at my computer during the early stages of this project. Your place on the desk is always reserved. You are missed but not forgotten. Fast Eddie, Truman, Harper, and Peanut did their best to get me out of my own head (and to feed them or take them for walks) whenever they could.

Finally, I owe the biggest thanks of all to my partners in crime, Nancy, David, and Trinity. Nancy Nichols-Pethick has lived with this project every bit as much as I have over the years and I could not have finished it without her: without her keen intellect, without her love of the small glories in life, without her flights of fancy, without her profoundly good nature, without her infectious laugh (complete with snort!), without her slightly naughty sense of humor, without her compassion, without her patience, without her understanding of what makes me tick.

David and Trinity…well, you two changed everything…and all for the better. You make it all worthwhile.

1

INTRODUCTION

A Knock at the Door

If the current television schedule is any indication, Americans are fascinated with images of crime and punishment and becoming more so every day. On any given evening, viewers can catch a glimpse of their tax dollars at work as both real and fictional police struggle to maintain an increasingly tenuous order on the streets of our cities. And what is perhaps most striking about the rise of interest in these images is the variety of formats that have defined the form over the last three decades: from the *video-vérité* format of *COPS* and *Police POV*, to re-enactment formats such as *America's Most Wanted* and *Rescue 911*, to human interest news magazines such as *48 Hours*, *20/20* or *Dateline*, to the emergence of Court TV, the cable network devoted solely to stories and news about law and order in the U.S.[1] At the center of these proliferating images of crime fighting are the prime-time dramas that have been with us since the days of radio and that continue to prosper on the major broadcast networks and their cable counterparts. As I write in 2011, there are no fewer than ten police-related series on the networks' prime-time schedules and dozens more in syndication – from reruns of *Naked City* and *Police Story* on the Retro Television Network, to the repurposed episodes of Dick Wolf's NBC hit, *Law & Order: Special Victim's Unit* on the USA network. Cable networks have also made inroads into the genre by producing original police series such as Lifetime's *The Division*, TNT's *The Closer*, and FX's *The Shield*. Cable networks like USA and TNT have also taken over production of series, such as *Law & Order: Criminal Intent* and *Southland*, respectively, which were originally created for the broadcast networks. This era also marks the emergence of franchises built around brands such as *Law & Order*, *CSI*, and *NCIS*, which have delivered nearly exponential growth to television crime dramas.

Finally, many of these series (both past and present) extended and expanded life via the surge in DVD releases of entire seasons as well the viability of internet-based distribution models like Hulu, Amazon Instant Video, and Netflix.

Given the scope of even this abbreviated list of programs, it is puzzling to find that the history of police drama series in U.S. television has been told in deceptively simple terms – when it has been told at all. Anyone hoping to understand the history of the police drama on television and make sense of its continued popularity by turning to the work of critics and scholars will be disappointed. First, until recently there simply has not been very much written on the subject. Second, much of what has been written seems to repeat the same basic contention: that television police dramas are limited by the genre's formulaic nature. According to most critics, this formula provides moral reassurance and champions an inherently conservative social agenda by focusing on the essential wisdom and virtue of those who enforce the law (police officers, district attorneys, etc.) and offer protection from all who threaten the social order. Furthermore, the genre suffers a fundamental contradiction, in which its realist aesthetic (practical location shooting, handheld cameras, naturalistic lighting, etc.) is constantly undermined by the conventions of melodrama.[2]

Even the producers and writers who have been most responsible for pushing the boundaries of the police genre often express ambivalence about its potential for telling honest or innovative stories. The creators of the groundbreaking *Hill Street Blues*, Michael Kozoll and Steven Bochco, initially balked at the idea of writing another cop show when first approached by NBC executives in 1980. Kozoll later explained their initial reluctance: "Because no matter how well intentioned you are when you go out to do a cop show, it's almost impossible not to end up with a bag of shit afterward. Because we've all done those boring, heroic, tired, tired shows, and you're going to kill yourself and the public doesn't even want to watch them any more, and they really don't address a very serious issue" (Gitlin 2000: 280). Still, the debut of *Hill Street Blues* in January 1981 clearly signaled that this genre, which by most accounts had become hopelessly mired in cliché, was capable of delivering something more profound and, under certain conditions, extraordinary. Certainly, like any television genre, the police drama series is a product of formula, made in a commercial system that depends on repetition and reliability. But what *Hill Street Blues* demonstrated was that these conditions do not entirely foreclose the possibility of significant innovations.

The world of *Hill Street Blues* was no weekly palliative; it was a dark, foreboding, and anonymous American city under siege. The spectacle of harsh poverty and abandonment stood as the uncompromising background for a theater of desolation in which the primary players were a precinct of world-weary cops defined as much by their flaws as by their virtues. Their antagonists came not only from the streets – ethnic gangs, drug addicts, pushers, pimps, and prostitutes – but also from within the halls of city government and a failing criminal justice system. Storylines proliferated from every shabby corner, tangling into one another and splitting apart, sometimes becoming resolved, sometimes not. The settings were bleak and the action was chaotic; the humor was dark and the drama sobering. And the police genre, so the story goes, was forever changed.

But this tale of divine salvation is also too simple. *Hill Street Blues* did not spring fully formed from the minds of its creators. Like most television dramas, it came to life in an intricate web of intentions, accidents, and compromises within the setting of commercial television, where creative vision runs up against industry economics, network programming strategies, and the technical challenges of production.[3] Because *Hill Street Blues* emerged from this complex institutional framework, a number of critics have viewed it as a miraculous *exception* to the logic of a monolithic "system that cranks out mind candy" (Gitlin 2000: 273).[4] In fact, the opposite is true. The producers of *Hill Street Blues* operated within familiar frameworks provided by the television industry and the conventions of the police drama to expand the range of what could be represented and discussed in the genre. Genres exist not only as broad categories for connecting and differentiating texts, but also as a set of recognizable stories, plots, settings, and characters that can provide a foundation upon which a range of topics may be addressed. *Hill Street Blues* used the familiar confines of the station house and the familiar crisis–response structure of police dramas as a way of openly (and often humorously) addressing the agonies and frustrations of modern day police work – the real human frailty beneath the guise of heroic toughness – and the real, though often dangerous or irrational, fears expressed by communities in the face of increasing crime.[5] In short, *Hill Street Blues* demonstrated particularly well what critic John Corner has called the "extraordinary cultural dynamics" of television: the medium's ability simultaneously to "ingest" elements of the "culture-at-large" and to "project" those elements back into the cultural arena in the form of stories (1999: 5).

The writers and producers of *Hill Street Blues* may have been particularly adept at managing the cultural dynamics described by Corner, but subsequent police series have followed suit by explicitly challenging or expanding the genre: *Homicide, Law & Order, Miami Vice, Cagney and Lacey, NYPD Blue, The Shield, The Wire*, and *CSI*, to name just a few. Each of these series, in fact, calls attention to the ways in which all of television's representations of cops perform this cultural dynamic. Perhaps more than any other genre, the police series in all its manifestations embodies the cultural dynamics of television in the way it necessarily incorporates (ingests) a range of current social and political issues (in the name of timely realism and relevance), filters them through the "rules" of the genre, the demands of coherent narrative, and the structure of the industry itself, and then projects them in the form of stories about individuals (police, lawyers, criminals, victims, witnesses, citizens) caught in the messy webs of politics and power. Stories about the police are more than ritual struggles between good and evil or reminders of the ultimate necessity of a benevolent police authority. They respond to some of our most pressing social concerns: concerns about how we imagine and maintain a sense of community in a vast and often alienating society, and about our rights and responsibilities as citizens. These stories play a vital role in the way we understand and engage issues of social order that most of us otherwise experience only in such abstractions as laws and crime statistics.[6] And given the current diffusion and popularity of the form, we might ask a number of questions that deserve serious critical attention: Under what circumstances have stories about the police proliferated in popular culture? What function do these stories serve for both the television industry and its audiences? Why have these stories become so commercially viable for the television industry in particular? How do stories about the police help us understand current social and political debates about crime, about the communities we live in, and about our identities as citizens?

By way of delimiting the scope of this study, I want to focus on the last three decades in the history of television police drama: from the debut of *Hill Street Blues* in 1981 to the end of the 2010–2011 television season. This particular period comprises an especially important moment for studying the television police drama. First, this period has seen three significant developments in the *style* of the police series. With the introduction of *Hill Street Blues* in 1981, the police genre began to embrace a range of new narrative and visual strategies: most notably, increasingly "cinematic" camerawork, serialization of narrative threads, and the use of ensemble casts.[7] This period has also been marked by the increasing importance of generic hybridity. *Hill Street Blues*, along with *Miami Vice* and *Cagney and Lacey*, for example, often combined stories about police with narrative or visual strategies borrowed from other sources: soap-opera, MTV, and the Movie-of-the-Week, respectively.[8] At the same time, this period saw the emergence of "reality" television. Reality series trade on the aesthetics of immediacy borrowed from both the *cinéma vérité* tradition of documentary filmmaking and the exposé format of network news magazines such as ABC's *20/20* or NBC's *Dateline*. Some of the first successful examples of this format include Fox's *COPS* and *America's Most Wanted*.

Overtly cinematic camerawork, increased hybridity, and the rise of reality television highlight what John Caldwell has called the "excess" that is "the symptom of a much broader period of transition in the mass media and American culture" (1995: 5). These stylistic developments are, on one hand, purely aesthetic choices based on the need to differentiate each series from others like it; but they also serve as important elements in the marketing and programming strategies of the networks in the face of industry expansion and heightened competition for audiences that are segmented along increasingly specific socio-economic lines.[9] It is no coincidence that *COPS* and *America's Most Wanted* were the products of Fox, the first new broadcast network to appear on American television since the 1950s. The appearance of the Fox network signaled a new level of economic competition in the television industry and exposed the first chink in the armor of the big three networks that had dominated American television for three decades.

The years 1981–2011, then, represent a period of enormous change in the television industry. This period has seen both profound technological expansion (e.g. VCRs, cable, satellite, and internet distribution) and sweeping deregulation of the industry (e.g. the Telecommunications Act of 1996 and unprecedented consolidation of ownership). American television is no longer predicated on the model of the three major networks, each competing for the largest share of the national viewing audience. Over the past three decades, this model of television (variously referred to as the "Network Era" or the "High-Network Era") has been challenged by technological innovations such as VCRs, cable networks, and satellite and web-based delivery services that have fragmented the viewing audience across lines of race, ethnicity, gender, income, and age.[10] Concomitant with these technological developments, the business of television itself has undergone massive reconstructive surgery at the hands of a legislative agenda predicated on the laissez-faire principles of the marketplace. The result is that we have moved into what Michael Curtin has called the "Neo-Network Era" and what others have labeled the "Post-Network Era."[11] This new structure is based primarily on the dispersal of the "unified" audience into a range of smaller consumer groups, each with specialized interests and desires, and on program formats designed to meet those desires. Given these changes in the way the industry responds to and constructs the relationship between its audiences and its products, then, how might we re-think our understanding of the cultural function of the police genre? Within a changing institutional environment, what kinds of stories about the police find purchase? For whom are these stories being told? How can we move away from a model of the genre that ritually underscores the discovery of a conservative ideology, and toward a more flexible model that accounts for the variety and popularity of the genre across an expanding television landscape?

One way to begin addressing these questions is by paying more attention to the effects that the structure of commercial television production and distribution have had on the "cultural dynamics" of representation. The process of ingestion and projection is selective (not everything captures the imagination to the same degree or seems equally worthy of our attention), and as institutional dynamics change, and stylistic developments accrue, the performance of this dynamic changes as well. In the post-network era, different television networks diverge in their responses to broader social issues (such as crime) depending on their perception of their audience (e.g. Oxygen's target demographic is primarily professional women while Spike's aims at post-adolescent males) as well as the rise and fall of popular formats (such as crime-oriented documentaries on the Discovery Channel, The Learning Channel, and even Animal Planet).[12]

Finally, within this expanded institutional context, the police genre also needs to be understood as responding to (and sometimes *not* responding to) the changing social conditions of the period and discourses about these conditions. The rise to prominence of the "New Right" signaled by Ronald Reagan's presidential victory in 1980 emphasized increased levels of incarceration and more severe punishment, coinciding with an increased emphasis on "street crimes" (as opposed to "white-collar crime").[13] Similarly, the economic decline of inner cities and the related rise in "street gangs" in the late 1970s and early 1980s fueled the "war on drugs" which has, in turn, led to an increasingly overcrowded prison system.[14] In terms of the police themselves, post-*Miranda* criticisms of the rights of the accused and the obligations of due process sit alongside increased media attention to accusations of police brutality and corruption, particularly in the wake of the Rodney King beating, the Rampart precinct scandal in Los Angeles and the Abner Louima case in New York. Additionally, the past two decades have seen the rise of the victims' rights movement that has placed special emphasis on domestic violence, sexual offenses against children, and the right to personal and community notice regarding the release of offenders.[15] The concern with community notification about sex offenders sits alongside the increasing popularity of broad-based "community policing" initiatives that have grown steadily since the 1970s. This period includes an increasingly fervent anti-American nationalism arising from economic globalization, religious fundamentalism, and separatist movements within the U.S. And, finally, the period is indelibly marked by the memory of September 11, 2001, which has provided the shared backstory for almost every police drama since that date.[16] These conditions (and more) form part of the cultural dynamics of the genre. The issues supply the police drama with a sense of relevance and immediacy, but they are always filtered through the genre, molded by production practices and industry regulations (regarding violence and language, for instance), and presented as stories to audiences. Through these three overlapping contexts – textual, industrial, and social – this study explores the specific material conditions within which a particular cycle of the television police genre emerged and developed between 1981 and 2011. The central goal is both to historicize the police genre and to demonstrate that it is a far more flexible, varied, and dynamic form of cultural production than has previously been acknowledged.

Getting In: Critical and Textual Contexts

Scholarly criticism of the television police drama has had a tendency to approach the police series as an inherently conservative form that repeatedly rehearses a dominant "common-sense" ideology of law and order. This vein of criticism often focuses on "traditional" series, which are structured as morality tales, with a heroic figure or team setting out each week to bring criminals to justice. On the other hand, there also has been a tendency to celebrate a few series as part of a broader spectrum of "quality television," in which series are marked by multiple narrative threads, complex criminal motives, and often introduce mitigating circumstances that complicate issues of guilt and innocence. In what follows, I explore in more detail these two threads of criticism as particular discursive strategies for making cultural distinctions and evaluations within the police genre. I then place these critical approaches within the broader contexts of genre criticism, dialogic theories of language, and ideological analysis, and then suggest some of the ways we might deal more adequately with representations of crime and punishment as a particular arena of cultural production.

The cycle of "quality" police dramas that have followed in the wake of *Hill Street Blues* – critical successes like *Homicide: Life on the Street*, *NYPD Blue*, and more recently *The Shield*, *The Wire*, and even short-lived commercial failures like *Brooklyn South* and *EZ Streets* – continue to build on *Hill Street*'s formula: ensemble casts with anti-heroic cops, gritty urban settings, dark humor, a visual style that approximates a version of documentary realism, ongoing serial narratives, and a willingness to let loose ends of the plot dangle. Closely connected to these more ambitious narratives are series such as *Law & Order*, *Cagney and Lacey*, and *Miami Vice*. While the action of these series usually focuses on one major case for each episode (and resolves that case by the end of the hour), they offer complex characters who are often haunted by their pasts or by the ethical dilemmas of policing, and who have changed over time. These series, along with a range of other "quality" programs of the 1980s and 1990s, such as *LA Law*, *Moonlighting*, *China Beach*, *Picket Fences*, *ER*, and *Northern Exposure* (to name only a few), have been celebrated as comprising a uniquely artistic period in American television: a renaissance or second "golden age."[17]

But at the same time, more "traditional" series such as *Hunter*, *Nash Bridges*, *In The Heat of the Night*, *T.J. Hooker*, and the more recent *CSI* and *NCIS* franchises, have also enjoyed continued popular appeal, often scoring higher ratings than their more prestigious counterparts.[18] These more "traditional" police series typically follow a closed episodic structure in which all major narrative threads (usually one particular case and a secondary narrative) are resolved by the end of each episode. These series tend to rely heavily on one of two primary formulas of the detective genre. The first we might call the "swift justice" model in which the perpetrator is known from the outset and the job of the police is to apprehend him or her. This form typically relies on heightened action sequences, especially car chases, fistfights, and gun battles, to bring closure to the narrative. The second structure is simply the "whodunit" model in which the criminal is unknown and the role of the police detective, using his or her superior sleuthing abilities, is to solve the mystery. Because these series rely so heavily on these recognizable structures and tend toward closure, they are often dismissed as trite and predictable escapism, sensationalist exploitation or, in the worst cases, politically reactionary dogma not to be taken seriously.

As I mentioned earlier, a great deal of the writing about the police genre to this point has advanced the argument that the structure and concerns of the genre are almost inescapably conservative.[19] Brooks Robards, for instance, states that the police series is "obsessed with coming out on top in the eternal struggle between right and wrong" and that this obsession means police dramas "comprise a deeply conservative genre that works to re-establish *over and over again* the importance of social order and the need for the elimination of wrongdoing" (1985: 11, emphasis added). Similarly, John Sumser argues that "there is a Social Darwinist tinge to prime-time mystery dramas that is used to justify the re-establishment of order at the end of each episode" (1996: 2). Elsewhere, B. Keith Crew argues that the "very nature of TV crime dramas ... causes them to reproduce the myth of crime and punishment at the expense of other myths with more critical and progressive orientations" (1990: 132). This myth of crime and punishment to which Crew refers is "almost inevitably conservative" (1990: 131).

In many cases, the "conservative" thesis relies on the recognition of the unquestionable moral righteousness of the police. This righteousness is represented as either a distanced and paternal professionalism (as in *Dragnet* or *Adam-12*) which is always successful and in which guilt is absolute, or as a heroic willingness to risk life and livelihood in pursuit of justice – even if it means working outside the boundaries of the institutionalized justice system where police procedure can turn into vigilantism. In the latter model, even in cases where the police fail in their efforts to apprehend or retain a criminal within the boundaries of "proper procedure," the police series is seen as lodging a reactionary complaint against a *system* of law and order which is too "soft" or "liberal" – too concerned with the rights of the accused to be effective. In either case (but especially the latter), police violence is mitigated as being "legitimized by its objects – those outside the consensus, deviants, or ... villains" (Dennington and Tulloch 1976: 39–40). As Stuart Kaminsky argues, the heroic structure of the police story justifies every decision the officer makes: "In a continuing series, the myth demands that whatever the officer decides will be made right" (1985: 58). Similarly, Crew suggests that cop shows typically divulge the criminal's identity early on (as opposed to mysteries which withhold the criminal's identity) in order to emphasize the guilt of the criminal and to focus the drama on the process of apprehending him or her. The knowledge that "the bad guys are guilty excuses illegal behavior on the part of the TV police" (Crew 1990: 133).

Interestingly, claims for the *exceptionality* of the "quality" series from *Hill Street Blues* through to *The Wire* often rely on these same kinds of formal features, although the terms are typically reversed. Predictable narrative coherence turns to celebrated fragmentation; moral certainty fades in the light of the hero's complicated or troubled personal life; the mythical tidiness of a job well done is transformed into a more realistic "messiness" of cases that cannot be solved. The immediate reactions to *Hill Street Blues* provide perhaps the most far-reaching and illuminating examples of this kind of criticism. From arguments outlining the modernist sensibilities of the fragmented *Hill Street Blues* narrative (Deming 1985; Thompson 1996), to a range of cases made for the "metaphoric vision" of the series (Landrum 1984; Castro 1985; Zynda 1986; Ziegler 1987) to Joyce Carol Oates's famous *TV Guide* paean to this "Dickensian" series as the one program that she and her Princeton colleagues could all enjoy (1985: 4), to a wide range of reports on the maverick, tradition-bucking spirit from which the show emerged

(Gitlin 2000; Hanson 1984; Kerr 1984; Stempel 1992; Thompson 1996), most of the writing about *Hill Street Blues* shares one central thesis: this was no *ordinary* cop show. In fact, many of the claims to exceptionality seem to stem from a desire to actually *remove* the series from the police genre by repositioning it as a generic hybrid or something else all together. *Hill Street Blues* was seen largely as a soap-opera melodrama about cops, a form receiving a great deal of critical attention at the time (likewise, *Miami Vice* was the "legacy of film noir" and *Cagney and Lacey* was often considered an explicitly feminist series about "two women who just happened to be cops").[20] In making these moves away from the generic category of the police series, and by carving out a space for particular "quality" series as something different, these celebrations of exceptions only underscore a more deeply held belief in the otherwise "ordinary" conservatism of the genre.

In each of the cases described above, the "conservative" thrust of the genre seems to be located in the inevitable movement of the "typical" or "traditional" narrative toward an ultimate resolution of the criminal plot: the obviously guilty party is either dead or behind bars and the hero is reconnected to the "Law" by way of his or her moral victory, ready for next week's adventures.[21] In the exceptional series, comforting resolutions are often replaced by what Todd Gitlin labels a "series of holding actions" – a sense of defeat indicative of a neo-conservative, social-Darwinist turn in social policies toward crime in the 1970s and 1980s (2000: 315). What all of these critiques seem to share is a desire to illustrate that the formula is basically conservative, even reactionary, in its reproduction of the status quo – either through the celebration of heroic action or the acknowledgement of futility.

Instead of focusing our attention on the question of whether or not the police series is conservative, however, I see these critical positions as discursive strategies for understanding and defining, interpreting, and evaluating the police genre.[22] This discursive approach has been central to genre theory in recent years and provides a useful way to move beyond generally repetitive ideological analyses of generic texts and to think about the discursive work of generic categories themselves. As Jason Mittell has argued, genres are not simply components of texts but, rather, they are larger cultural categories, defined in the intersections of production, reception, and critical practices in specific historical contexts (2001: 7). Texts contain generic identities, but do not constitute the genre. Instead, "genre" as a concept also circulates outside of texts – as a way of positioning, understanding, and evaluating individual texts or groups of texts. Mittell's arguments fall in line with Rick Altman and Steve Neale who both suggest that genres are produced in the discourses surrounding texts and the multiple and sometimes conflicting *uses* to which texts are put.[23] Mittell labels this arena of use as the "cultural life" of a genre, indicating that generic labels are most often used to privilege some meanings over others (e.g. emotional supportiveness versus heroic individualism), or to position texts in different ways for different audiences (e.g. melodrama versus police procedural). What this discursive approach offers is a way to deal with genre as a concept without always returning to textual examples in the last instance – a practice that often leads critics to argue for a particular "correct" reading of a text, or to emphasize the maturing process or "progression" of a genre through its increasingly self-referential texts.[24]

While I want to take up this discursive argument in relation to the television police series, one danger that this approach presents is that it has the potential to lose sight of the actual texts themselves (though this is not a mistake that Mittell makes). Rather than regard the text as something that enters into discourse *after the fact* and that is to be *acted upon* by various user groups, I want to consider the discursive character of the texts themselves: the way certain generic tropes (themes, characters, settings, situations) are activated as part of an ongoing cultural *dialogue* about the social experience of crime. On one hand, the social experience of crime is caught up in the dynamics of television representations of the police (one of the ways we experience crime is through the stories that we see and hear and read in the media). On the other hand, these generic texts are discursive commentaries on the genre in their own right. Critics and marketing executives are not the only people concerned with generic meanings: producers of police dramas also engage the genre directly in their need to deliver simultaneously recognizable and differentiated texts. As Horace Newcomb and Paul Hirsch have stated: "The goal of every producer is to create the difference that makes a difference, to maintain an audience with sufficient reference to the known and recognized, but to move ahead into something that distinguishes his [sic] show for the program buyer, the scheduler, and most importantly, for the mass audience" (1983: 510). Every example of the police procedural, then, must be seen as engaging in a dialogue not only with larger social currents but with other examples of the genre and institutional developments as well.

This notion of "dialogue" stems from my own interest in the literary, linguistic, and social theories of Bakhtin and Voloshinov and Stuart Hall's related concept of "articulation." In particular, I am interested in pursuing Bakhtin's ideas about language systems and "heteroglossia" – the idea that many languages are "embedded in any given social language" (Newcomb 1984: 39) – and how this heteroglossia disrupts notions of ideological rigidity. At the risk of oversimplifying these ideas, Bakhtin was centrally interested in the way the ideological force of a "unifying language" (such as a national language or a generic category) always operates "in the midst of heteroglossia" (1994: 75). For Bakhtin, what was important was the way in which the "unifying language" is always made up of different dialects and accents, different ways of using and inflecting the language and the meanings of words from a particular social position. Every utterance, then, is born into a dialogue within the larger language structure:

> Indeed, any concrete discourse (utterance) finds the object at which it was directed already as it were overlain with qualifications, open to dispute, charged with value, already enveloped in an obscuring mist – or, on the

> contrary, by the "light" of alien words that have already been spoken about it. It is entangled, shot through with shared thoughts, points of view, alien value judgments and accents.
>
> *(1994: 75)*

In other words, the meaning of a specific "utterance" is not guaranteed by the fact that it has been "spoken." Instead, an utterance is meaningful only in the way it is articulated to other utterances around it and to the social forces that comprise what Stuart Hall calls "ideological terrains of struggle" (2003: 41). These terrains are determined by "the existing balance of social forces" found within specific historical and material contexts (2003: 45). The reductive concept of a fixed ideological meaning is replaced by a theory of "articulation" which insists on replacing absolute ideological determinism with a "theory of contexts" (Grossberg 1993: 4). As Hall has argued: "The 'unity' which matters is a linkage between the articulated discourse and the social forces with which it can, under certain historical conditions, but need not necessarily, be connected" (2003: 53).

In thinking specifically about television police dramas, however, we need to be careful about relying too much on an analogy to speech acts or literary texts. We need to consider television series as "utterances" produced within and by an immense commercial industry that operates according to unique and specific economic imperatives (different from – though related to – those of other media industries). But conceptualizing a popular commercial genre like the police series as a kind of unifying language system (or produced within such a system) does not necessarily mean that its utterances (series, episodes, plots, scenes, shots, etc.) are tied to a guaranteed set of meanings. As such, I am interested in considering the police genre not as a product of a unified (and unifying) language of conservatism, but as built of concrete utterances that enter into the "charged" and "disputed" dialogue about crime and punishment in America in the late 20th and early 21st centuries. The meaning or ideological thrust of a series like *Hill Street Blues* or *T.J. Hooker*, for example, first must be recognized as always conditional – never guaranteed – and then related back to the particular ways that it articulates a set of social discourses about crime and the work of policing already circulating in the culture. Seen in this way, television representations of the police can be understood as relying on and engaging with a range of competing perspectives on crime and punishment in order to tell stories that matter. Rather than compulsive repetitions of "dominant ideology," they are dynamic participants in a broader "cultural forum" or "terrain of struggle."

At this point I would like to offer a few brief examples of relatively recent thinking about crime and popular culture that seem to make these broader connections in useful ways. These examples each come from outside the field of television studies; they come from history, literary studies, American studies, and sociology, and offer ways to think about television representations of the police as articulated with broader historical and cultural processes that shape how we come to "know" crime (to understand and recognize it) within specific material and social conditions.

In *Murder Most Foul*, a study of the rise of the Gothic tale of murder, Karen Halttunen locates a key break in the historical treatment of murder as a transgressive act that occurs in the late 18th century with the first murmurings of Enlightenment Liberalism. The crime of murder was once understood through the lens of execution sermons, in which the murderer was established as a "common sinner" – a debased example of the moral affliction that threatened the entire community. These sermons were public performances, but also circulated in printed form as a kind of early popular crime literature. As Halttunen argues, "the story of murder officially entered print within the framework of the execution sermon" which "shaped the official narrative of the event along highly formulaic lines" (1998: 8). The purpose of these sermons was transcendence: they were designed to "remove attention from the murderer's past transgression and focus it on the future state of his soul" (1998: 31). But during the late 17th and early 18th centuries, crime was gradually placed into a new framework: one that emphasized the Gothic conventions of horror and mystery, and identified murder as a monstrous act for which there was no adequate explanation. Importantly, these secular accounts circulated as trial reports and emphasized aspects of the crime that differed from those in the sermons, redirecting the readers' attention "from the spiritual destiny of the convicted murderer to the crime itself, in all its bloody violence" (1998: 32). These trial reports were, in many ways, prototypes for popular sensationalist literature and reporting in an increasingly literate society that began to view literature as a commodity to be consumed quickly. What made these trial reports so important was that they "endeavored to replace the sacred narrative with a new mode of coming to terms with crime" at a time when the cultural authority of the clergy was waning (1998: 3). This notion of "coming to terms" underscores the idea that crime is not a neutral or natural set of categories, but a set of human constructs. These constructs are "the mental and emotional strategies employed within a given historical culture for responding to serious transgression in its midst" and act as one means by which we come to interpret human nature (1998: 2).

Halttunen presents us with a model for thinking about how and why changes occur in the conventions of storytelling about crime and punishment and in the cultural authority to "make sense" of crime. This model addresses attitudes and beliefs about what constitutes crime and how punishment is justified, and how these attitudes change over time. The model also addresses what kinds of historical conditions (such as literacy) produce changes in discourse, and how these discourses, in turn, respond to and help shape our experience of these conditions. Finally, Halttunen's model points to larger questions about how the dynamics of popular culture – and genres specifically – help to constitute and also to contest the cultural authority to make sense of larger social issues. What Halttunen offers to the study of contemporary police dramas, then, is a clearer sense of how changes in the form and function of crime stories are not simply matters of building a "better" story about crime – be it a 17th century execution sermon or a 20th century police series – but how these changes are indicative of larger shifts in cultural authority and changing conditions of consumption.

This issue of cultural authority is absolutely central to Christopher Wilson's *Cop Knowledge*, an account of police authority in 20th century America. As Wilson argues, police authority is also a kind of narrative authority. In other words, Wilson explores "the role of police power in cultural storytelling" within a range of different production contexts: the early social criticism of Stephen Crane and Lincoln Steffans; the crime novels of Joseph Wambaugh; the true-crime reporting of David Simon and Jimmy Breslin; and the New Right criminology of James Q. Wilson and George Kelling (2000: 6). For Wilson, the police have a fundamental and authoritative role in shaping popular representations and interpretations of crime and the strategies to prevent crime. Not only do the police do *interpretive* work – responding to everyday calls for help and then determining whether or not a crime has even been committed – they are also active *producers* of public information.[25] As sources of information about crime and police procedure (for politicians, social scientists, reporters, and television producers eager to "get it right"), the police often offer the "first drafts of a great deal of our cultural knowledge about social disorder and criminality" (2000: 6). Those figures that rely on the police for information, in turn, reproduce the figure of the police within a range of conventions typically predicated on the "working cop" with "an intimate, interstitial, nearly mystical understanding of crime, urban neighborhoods, working-class identity, 'fallen' political citizenship, and more" (2000: 215). For the study of television police series, Wilson offers a way of thinking about the character of the police officer as a complex cultural construction – as both a source and a product of popular representations of crime. Rather than simply being the ultimate arbiter of law and order, the figure of the cop is an interstitial figure, the product of a range of sometimes contradictory discourses: at once the figure of a distanced and paternal authority (one who cannot "take it personally"), an involved social worker, and a holder of a special, "mystical," class-based knowledge about "the way things are." A key question for thinking about television crime drama is how this complex figure of the police officer negotiates between different layers of society. How does the character of the police officer "speak" to different audiences and how have representations of the police shifted to meet the changing demands of commercial television? How is "cop knowledge" and authority structured into police series in ways that both fuel and soothe cultural anxieties?

While Wilson and Halttunen address crime and its representation in broad historical contexts, Richard Sparks offers perhaps one of the first and most comprehensive accounts of the social function of the police series on commercial television. Taking up questions about cultural anxiety and moral panic in relation to television crime fiction, Sparks suggests that "crime fiction presupposes an inherent tension between anxiety and reassurance and that this constitutes a significant source of its appeal to the viewer" (1992: 120). While the resolution of this anxiety is, for Sparks, the probable location of the ideological thrust of police stories, he is less interested in the "falseness" of the reassurance than with the question of why we should take pleasure in these stories in the first place and how this pleasure "stands in relation to our everyday conditions of life" (1992: 120). Sparks argues that previous ideas about media anxiety presuppose a stable "effect" (the creation of fear) without really considering the different social and discursive positions that different viewers occupy. As Sparks states, "there is simply no morally or factually neutral language in which to narrate stories about crime and punishment" (1992: 16–17). Thus, we need to understand television representations of the police as bound up with larger "cultural and political matrices" and the fear of crime as "itself only intelligible when it is understood in terms of the position of crime as a discursive arena of public life" (1992: 17). Rather than looking simply at the kinds of representations contained in police series (the realm of content analysis), we need to look outward to public debates that help frame our understanding of crime and also help shape media representations of how the police deal with crime. For instance, in terms of police series in the 1980s and 1990s, we need to come to terms with issues such as the rise in public discussions about community policing, victims' rights movements, the redesigning of urban spaces (the most often represented location of violent crime), and police scandals and reform. These issues are hardly neutral: they are shot through with contradictions and disagreements arising from the experiences of citizens in different social positions.

These three examples connect in their ability to consider popular representations of crime as not only stories *about* crime, but as part of the dialogic process by which the public comes to know, understand, and react to crime. Each of these critics sees stories about crime and the police as articulated with particular discourses: the mystery and horror of crime (especially murder), the cultural authority of the police officer, and the discourse of fear and moral panic in modern societies. These discourses circulate through most television police stories in one form or another. But rather than seeing them as simple generic markers that deliver a narrow range of meanings, these critics all underscore the importance of considering these discourses within larger cultural debates about what constitutes crime, how we understand it, and how we live with it.

Canvassing the Neighborhood: Discourses of Crime, Community, and Citizenship

The critical writing about television police drama offers a useful overview of the central narrative concerns of the genre and its formal elements, but there has been too much emphasis placed on finding similarities – the common plot elements, the primary themes that "define" the genre – which often leads to direct connections between form and ideology. This focus on uniformity in genre criticism is, admittedly, a realistic response to the vast quantity of material that faces any critic of television or popular culture more generally. Perhaps because of this glut of material more ambitious projects, such as Christopher Wilson's, tend to ignore television (often giving it just an occasional nod) in favor of literary or cinematic texts. It is my contention that, because of the wide range of styles and subjects available on an increasing number of programing outlets, television may be one of the richest cultural arenas for investigating representations of crime and policing.[26]

The idea of an arena of multiple and competing perspectives within the context of commercial television is argued most directly in Horace Newcomb and Paul Hirsch's article, "Television as A Cultural Forum," which has had a significant impact on my own approach to thinking about the police genre. Newcomb and Hirsch are concerned with the arena of ideological *conflict* and *contradictions* embedded within television texts. They argue that "these contradictions locate, identify, construct, and contextualize conflicts and contradictions in society at large" (1983: 500). In general, Newcomb and Hirsch are arguing for an approach to commercial television that understands television texts not as simple transmitters of ideology, but as complex artifacts that are deeply implicated in an equally complex process of cultural examination and interpretation:

> In its role as central cultural medium [television] presents a multiplicity of meanings rather than a monolithic dominant point of view. It often focuses on our most prevalent concerns, our deepest dilemmas. Our most traditional views, those that are repressive and reactionary, as well as those that are subversive and emancipatory, are upheld, examined, maintained, and transformed. The emphasis is on process rather than product, on discussion rather than indoctrination, on contradiction and confusion rather than coherence.
>
> *(1983: 506)*

Newcomb and Hirsch are not so concerned with the outcomes of narratives – the tidy resolutions that we are all so used to – but, rather, the paths that narratives take in getting to their resolutions: the twists and bumps and forks in the road, the process of examination. They point out that "in popular culture generally, in television specifically, the raising of questions is as important as the answering of them" (1983: 507).

Of course, Newcomb and Hirsch have been criticized for constructing a model for television criticism that is perhaps too open and pluralistic, and, therefore, fails to account for the unequal distribution of power in society. Discourse must be understood as existing within very real and material power relationships – in dominant value systems and conventions for interpreting experience.[27] As John Fiske has argued: "The way that experience, and the events that constitute it, is put into discourse – that is, the way it is made to make sense – is never determined by the nature of the experience itself, but always by the social power to give it one set of meanings rather than another" (1994: 4). For a number of critics interested in exposing the ideological work of television, any questions raised within television programs are contained by the "hegemonic process" of television representation. The emphasis of this type of criticism is on the *resolution* of contradictions. Todd Gitlin, for instance, has argued that "major social conflicts are transported *into* the cultural system, where the hegemonic process frames them, form and content both, into compatibility with dominant systems of meaning" (1979: 532, emphasis in original). While Gitlin, along with Newcomb and Hirsch, is certainly concerned with the "cultural dynamics" of television, for him the hegemonic process is totalizing and already embedded in formulas, genres, character types, settings, "slant," and narrative solutions which all work to delimit the range of interpretations that are possible (1979: 519).[28]

But what makes the idea of a "cultural forum" or "arena" model useful for thinking about the police genre since 1981 is precisely that it does not short-circuit analysis by presuming an underlying unity of discourse enforced by the institution of commercial television. In refusing to see the "meanings" of all television texts as ultimately and always determined by their status as industrial formulae, this model helps account for the mounting variety and difference that marks commercial television today; and it does this in ways that do not simply see this variety as an ideological subterfuge. The idea of the "cultural forum" more adequately accounts for the experience of television in the "post-network era" (though this argument may seem counter-intuitive at first).[29] The experience of television is defined more and more by niche audiences and programming that is designed to attract those viewers defined as demographically desirable while rebuffing others – a situation in which the only dialogue going on would seem to be a kind of "preaching to the choir."[30]

Within this expanding television landscape, however, competing networks strive to differentiate themselves from one another and build a loyal and steady audience base; one way to do this is by placing generic texts within new contexts – to make them speak in a slightly different accent. In this way, genres themselves become a location of difference and variety. For example, there are not only police series on the major networks (which still adhere, at least in part, to the practice of trying to attract a general audience), but also on cable networks with a much more narrowly defined and marketed audience. Because of these distinctions within the audiences, a police series produced for Lifetime (such as *The Division*) is likely to differ significantly from a series made for FX (such as *The Shield*).[31] This expansion does not signal a breakdown of the "forum" model. Rather, it *reconstitutes* it as a quality of genres.

Following this view enables us to think of television narrative and cultural discourse in ways suggested by Halttunen and Wilson, and to think of the police genre not so much as a rationalized system of conventions and formulas working toward a common end, but as a site for dialogue, for engaging a variety of discourses about crime: what actions or behaviors constitute "crime" in the first place; how a criminal act is "known" as such; how crime is understood in relation to the establishment and maintenance of communities; how definitions of crime and community affect our notions of what it means to be a "citizen." In other words, the primary question, as Newcomb has stated elsewhere, is how to connect media texts – the police series in this case – to social experiences and practices (1984: 36). Rather than begin from the assumption that police shows rehearse a singular *a priori* conservative position towards crime and punishment, then, analysis should begin with the notion that no historical period offers a singular political discourse. Instead, the cultural artifacts of any period will reflect a rich cultural dialogue about particular subjects (such as crime); and this dialogue is manifested in both the repetition and variety inherent in generic texts. Analysis should also begin with questions about the cultural dynamics of television: ingestion and projection. How and why does television ingest certain issues (e.g. particular types of crime)? How and why are these issues inflected in certain ways by the demands of genre and by the institution of television? How and why are these issues reproduced and circulated, becoming part of the larger cultural process? Paying close attention to both the structure *and* variety of this dynamic will allow us to gain a more complete understanding of the range of social forces that exert pressure on cultural texts.

One of the central tenets of this study is that the political implications of the police procedural involve sorting through the interconnected discourses of crime, community, and citizenship. None of these terms has an objective status; each represents a category of competing values, and is constituted and made real through discourse. And in order to begin addressing the very real social and political effects of these terms, we need to address some central questions. How are crime, community, and citizenship understood as such? What determines why some acts and/or people are labeled criminal? Under what circumstances are crimes mitigated? What constitutes a community? How are laws interpreted in terms of community? How are issues of citizenship bound up with identity politics? How do legal definitions of citizenship differ from behavioral definitions? Only by addressing these sorts of questions can we begin to get a better sense of the discursive quality of these terms and how television representations engage with these discourses.

As a way of demonstrating briefly how a range of discourses about crime, community, and citizenship can operate in a text at any given time, I want to offer as an example a single scene from "Hill Street Station," the first episode of *Hill Street Blues*. In this episode, two officers, Hill and Renko, are called to handle a domestic dispute in an inner-city housing project. When the officers arrive on the scene, they find a woman wielding a knife and threatening her teenage daughter and her husband (the girl's stepfather). After some initial questions, it becomes clear that the girl and her stepfather have been sleeping together and the horrified mother has decided to take matters into her own hands. Renko responds in what must be understood as fairly standard police fashion: "What we have here is statutory rape, possession of a deadly weapon, intent to use a deadly weapon ... I think we ought to run 'em all in." Before Renko can begin the standard arrest procedure, however, Hill interrupts and tries to offer another solution. Instead of starting the wheels of justice that will only result in the break-up of this family, Hill constructs a peace agreement between the three family members: the daughter will stop dressing in plain sight of the stepfather; the stepfather will stop seducing the daughter; the mother will make herself more sexually available to her husband. Once it is clear that the threat of physical violence has subsided, Hill puts a final exclamation point on the exchange: "OK. This is the law in this house!"

The first thing that we should notice is the way that a range of voices and perspectives come into contact in this scene: the wife/mother, the husband/father, and the two police officers (importantly, the daughter is mostly silent and hidden away during the scene). Even the two officers voice different perspectives on the situation, though it could be argued that their two approaches are actually working together to first instill fear in order to take control of the situation (through Renko), and then to offer a solution that avoids arrest (through Hill). But we also need to recognize that not all the discourses are given equal weight in this exchange. The real authority in this situation lies with the police themselves and within a particularly patriarchal framework that renders both the (voiceless) daughter and the (hysterical) wife/mother as somehow causing the husband/father's indiscretions.

Within this hierarchy of discourses, however, a number of issues are raised with regard to how crime is handled in terms of community and citizenship, each of which connects with a larger discursive field. These issues are articulated to (and within) the framework of patriarchy and implicit sexism, but the point (following Stuart Hall) is that this connection is not absolute or necessarily guaranteed. The interpretive field here includes a range of interconnected discourses, sometimes competing and sometimes reinforcing one another. In terms of "crime," there exist both legal and situational discourses in this scene – one is understood through the letter of the law, the other through the (sexist) dynamics of family life. One of my central arguments is that "crime" is always subject to this tension between legal definitions and more experiential or circumstantial definitions. These competing definitions of crime are also played out in relation to ideas of "community" – how the interpretation of crime can vary according to the dynamics of a particular group. In this scene, the community is comprised of the family and Hill establishes "the law for *this house*." Similarly, much of the work of community policing efforts are predicated on the police and community members being able to establish and agree upon proper conduct within a community and, more importantly, to recognize who belongs and who does not. Stories about urban policing may have a different way of representing communities than stories set in rural or suburban locations (such as *In the Heat of the Night*). Finally, the discourse of "citizenship" rests on an ever-present tension between rights and responsibilities. Within the sexist framework of the family on *Hill Street Blues*, for example, the husband felt that he had certain (sexual) rights, but failed to recognize his moral responsibilities to his wife and step-daughter. Whether "citizenship" is considered a matter of legal (national) identity, or smaller, more provisional group identities, this tension between rights and responsibilities is always near the center of police dramas. How is one held accountable for one's actions – legally, morally, or culturally? How are legal definitions of crime manipulated in order to account for moral ambiguity? How are behavioral notions of "good citizenship" used to coerce citizens and communities into taking action? How are abject citizens (often involved in criminal activity) used to help the police in their efforts to solve crime? How do the police themselves negotiate their own roles as both citizens and representatives of the state?

Again, these three sets of discourses are intimately bound up with one another and cannot be separated easily. And as we move outward from focusing on one scene in one episode of one series, and begin considering the wide variety of different series in an increasing array of viewing contexts, we can start to get a sense for just how varied and complicated this genre is.

Any attempt at a cultural history of television requires a range of analytical frameworks. My own approach attempts to develop a way of thinking about the prime-time police series through a variety of related lenses: *institutional* (changing conceptions of genres and audiences in the American television industry), *sociological* (the intersection of social policy and police work), *representational* (the politics of difference), and *territorial* (the community as a discursive space). Given this structure, I offer a way of rethinking the police drama as a series of sometimes quite different, but interrelated, texts connected first by the material conditions of their place in a post-network context, and second by the larger social, cultural, and political questions about what constitutes a community and what constitutes citizenship within that community.

A range of questions shape the contours of this study. How do particular cultural forms (in this case, television police series) engage with and help us rehearse particular discourses about social life? More specifically, how do certain social and political discourses about crime, community, and citizenship get woven into fictional narratives about policing? How are political issues activated across a range of programs that deal ostensibly with the same subject? Another range of questions that is often displaced in these studies concerns how we might begin to account for the popularity and resonance of certain generic forms at particular historical moments. In other words, the popularity of the police genre on television has ebbed and flowed over the course of its history and formal questions alone cannot address these inconsistencies. Questions about formal features tend to be questions about cultural continuity and textual similarity (or evolution) rather than questions about specific historical periods of production and consumption. This project addresses these types of historical questions as a way of opening up the range of interpretive possibilities for police narratives as a whole and raising questions about how genres are constituted and made meaningful.

Chapter 2 explores how the industry has approached police dramas since 1981 in relation to drastic changes in production and distribution, as well as changes in the regulation of the industry by the federal government. The emergence and growth of cable networks and satellite and web-based delivery services has created a constant demand for programming to fill a growing schedule. This expanding playing field has also fragmented the audience across the channel grid and created a demand for immediately recognizable and engaging products. These changes have created a need for different approaches to conceiving of, and marketing to, these diverse viewers. In particular, this chapter considers these institutional changes in specific terms of production (genre-mixing) and distribution (marketing to the audience and, especially, syndication). I demonstrate that the police genre has proven to be an incredibly flexible form, almost infinitely renewable and malleable across a range of different contexts.

Chapter 3 examines the 1980s as a key period for adding to the language of the police drama and reconstituting the cultural forum about crime, community, and citizenship. The social contexts of the period encouraged producers and network executives to reconsider the traditional representation of the police as either paternalistic guardians or benevolent social workers. This chapter begins by highlighting some of the key social concerns of the decade: the decline of inner-cities, rights of the accused, victim's rights, community policing, and the war on drugs. Taken together, these concerns suggested that public attitudes about the function of the police in communities across America were changing and needed to be addressed by popular culture in new ways. Some of the earliest responses to these changes turned up in novels (especially those of Joseph Wambaugh) and films (*Dirty Harry*, *The French Connection*, *The Seven-Ups*, *Fort Apache*, *The Bronx*, *Escape From New York*), all of which portrayed a darker, more chaotic world in which flawed (that is to say *human*) police officers replaced the steady administrators of justice represented by Joe Friday in *Dragnet*. Television producers picked up on these trends and expanded the range of the police drama in several ways: narrative strategies (a focus on characters, the use of ensemble casts, and serial plots); expressive styles (from documentary realism to pop-art slickness); and thematic concerns (including ambivalence about authority and an increased emphasis on the personal lives of the police). Series such as *Hill Street Blues*, *Miami Vice*, and *Cagney and Lacey* pushed at the limits of the police drama in all of these ways. At the same time, however, more traditional representations of police, such as *T.J. Hooker* and *Hunter*, and even the stylistically innovative series, *COPS* prospered as well. This mix of series within a particular social and historical context supports two conclusions about the function of police dramas in popular culture: first, that the genre allows competing perspectives about key social issues (clustering around the categories of crime, community, and citizenship) across different series, creating a cultural forum; second, that our understanding of how genres develop and change must account for this *dialogic* function in place of a model of evolutionary progress. The arguments and conclusions in Chapter 3 form the foundation for understanding the police dramas of the 1990s that are taken up through three specific case studies in the following chapters. Each case study links aesthetic factors such as realism, melodrama, and narrative structure with the larger thematic/discursive issues of crime, community, and citizenship.

Chapter 4 examines NBC's *Homicide: Life on the Street* (1992–2000) as a stylistically self-conscious series. I demonstrate that the series played with the conventions of realism by heightening certain unconventional stylistic strategies (such as jump-cuts and overlapping action) in order to inject energy into the narrative without relying on clichés such as car chases and gun battles. I also argue that this attention to style is used as a way to comment on the genre itself, specifically its commitment to realism. Contained within the stylistic and narrative structure of *Homicide* is an acute awareness of the tension between reality and its representation on screen, and *Homicide*'s production design is often intended to call attention to the questionable claims to authenticity that police dramas so frequently rely on. Finally, the self-conscious and critical style of *Homicide* opens up a space for the series to engage with discourses about community, especially the idea that communities and their boundaries are symbolically constructed.[32] Filmed on location in Baltimore, *Homicide* regularly examines the boundaries – both racial and geographic – that divide the city and asks critical questions about how those boundaries are made meaningful – how they "make sense" – through both the work of policing and its representation in the media.

Chapter 5 looks at *NYPD Blue* (1993–2005) and argues that the series uses the heightened moral concerns of melodrama as a way to explore questions about what constitutes good or productive citizenship. Specifically, I explore how the melodramatic mode opens up questions about rights, responsibilities, and the consequences of our decisions and makes them central to the discourse about citizenship. *NYPD Blue* uses contemporary social discourses to explore the concept of virtue that is so central to melodrama. The series uses this melodramatic mode to construct an image of "intimate citizenship," based on interpersonal relationships and responsibility to oneself and to others – not just the state. Another crucial aspect of intimate citizenship is control. Just as the detectives work to contain social chaos by closing cases, the stories of *NYPD Blue* work to discipline its characters' emotional lives and teach them control in the face of personal crises. In the end, the series forges an image of the detective as the "model citizen": disciplined and self-reliant, but also open and empathetic.

Chapter 6 looks at the phenomenon of *Law & Order* (1990–2010) as an example of how television's cultural forum is a product not only of competing social discourses put in play by different series and/or episodes but of changing industrial conditions as well. The series uses a story-driven, episodic narrative structure with themes "pulled from the headlines," combined with a highly flexible ensemble cast, to open up multiple perspectives on the question of what constitutes crime and culpability. Though not all perspectives are presented equally, in keeping with the foundational idea of the cultural forum, the fact that competing opinions get aired at all is significant for a genre that has traditionally relied on moral certainty. The stability and flexibility of the *Law & Order* formula is key to understanding the series' eventual success in the syndication market and the subsequent rapid expansion of the franchise to include as many as four variations running concurrently. The chapter finishes by arguing that the *Law & Order* franchise itself comprises a cultural forum about crime, as the loosely connected series with different narrative foci bump up against one another on the prime-time broadcast schedule and in syndication.

Chapter 7 turns to the police dramas of the 21st century, as constituting a development in the cultural forum on crime, community, and citizenship. Through a general case study of three major series since 2000 – *The Shield*, *The Wire*, and *CSI* – this chapter points to the diffusion of the genre across multiple programming outlets: broadcast, ad-supported cable, and subscription-based cable. These distribution outlets operate under very different regulatory frameworks and thus open up new spaces for stories about crime. At the same time, the major trend in commercial television over the past several years is toward branded entertainment. Each of these series has been instrumental in establishing or extending a recognizable brand identity for its home network. Within these brand identities, the series go about their work telling stories about crime. This chapter provides a particular case study of the ways in which the three series address the issue of truth that is so central to the police genre as a whole, and how they relate the idea of truth to the problem of "corruption." By comparing the ways in which each of the series handles the problem of locating "truth" we can see the variety of approaches (stylistic, narrative, ideological) that has informed the genre, and how this range has increased along with the expansion of programming outlets. In the end this institutional expansion, along with the focus on branding, may force us to reconsider the very notions of genre that have animated this study.

Making the Case

One of the central points of this study is that criticism of the police genre as inherently conservative, or of certain series as exceptions to the rule, has worked to close down larger questions about how the police genre *in all its forms* functions as an important site of cultural production – as a place where our own knowledge about crime and justice is at least partially forged and where our deepest commitments to the meaning of citizenship and to the maintenance of our communities are put to the test on a weekly (if not nightly) basis. Above all, then, this study aims to open up these questions to include not only the most intelligent and provocative examples of the genre, but the most formulaic and predictable as well. The cycle of police dramas that has emerged between 1981 and 2011 cannot be explained away by romantic tales of a "second golden age" of television or of maverick writers and producers flying in the face of network demands. Nor can critical attention to this period be divided so neatly along aesthetic lines, as if production values equate with political or cultural values. Such views make the mistake of characterizing the police genre as a singular phenomenon: one that consistently reproduces a conspicuously uniform ideology, while a few anomalous rogues find a way to work around the system. And while it is noteworthy that a series like *Hill Street Blues* could be elevated to the status of "art," the aesthetic discourse surrounding the series has made it difficult to discuss it in any other terms. In other words, arguments for the *exceptionality* of *Hill Street Blues*, *Homicide*, or *The Wire* (against the blandness and predictability of *T.J. Hooker* or *Hunter*, or *NCIS*) have displaced larger questions about the "ordinary" creation, purpose, and use of police series on commercial network television.

More importantly, we need to understand not only how texts differ but also the material conditions that help explain this variety: the changing state of the television industry in which the shows are produced; the range of personal and professional relationships explored in the dramas themselves; competing notions of criminality and the related debates over the limits of policing that so often provide the fodder for these narratives; and the location of crime and police work in these stories – the communities where this activity takes place. In short, we need to be more attuned to the nuances of the historically specific economic and social conditions in which these texts are conceived, produced, and experienced and within which their popularity – their value as relevant and attractive cultural commodities – ebbs and flows.

As described earlier, the preceding thirty-year period in television history represents a major change in the make-up of the media industry in the U.S. Network dominance has been eroded by the incursion of cable, satellite and VCR technologies, computer games, and the internet; and programming outlets (traditional broadcast networks included) now aim at smaller segments of the audience. The police genre has been deeply affected by these changes. If stories about policing have traditionally dealt with the definition and maintenance of communities, then changes in how audiences (communities) are imagined by networks and producers require new narrative strategies, new ways of thinking about community, and new ways of knowing "crime." Within the narrowcasting logic of the post-network era, we need to ask not only what kinds of stories are being told but for whom.

This particular historical context has also seen significant changes in the social policies toward crime in the United States. At the center of these social policies are the key issues of race, class, and gender and the way certain bodies are marked as criminal. To "know" crime is to read through discourse; and as certain discourses are given the weight of policy they come to dominate as a form of "common sense." The police narratives of the post-network era are not only expressive of a certain range of social discourses, but can also tell us something about how we have constituted and maintained our communities over the last twenty years. These programs can offer us clues about how certain discourses attain dominant status while others, though allowed expression, remain marginal. As Julie D'Acci, in her analysis of *Cagney and Lacey* argues, "network television, its programs (or texts), its viewers, and its historical contexts are sites for the negotiation of numerous definitions and discourses, with certain ones achieving more power or 'discursive authority' at specific moments and for specific participants than others" (1994: 3).

Above all, this study offers some insights into the question of how cultural institutions such as television interpret social issues and circulate ideas about them. Just as institutions cannot totally co-ordinate and regulate consumption of their products, policing remains an incomplete and contingent activity, subject to reform and revision. Laws and genres live by similar rules, each reflecting agreed upon conventions of behavior and representation, or formulas of conduct. But texts, like laws, remain open to interpretation, and competing interests always enter the struggle for meaning, though never from totally equal bargaining positions. This struggle for meaning at the level of popular texts is an important political struggle that will necessarily leave its imprint on the future. The primary goal of this study, then, is quite simple: to put a name (and give a history) to the multiple representation of police work on American television. Like the detectives and beat-cop heroes of the stories, these texts do interpretive work and enter into a broader cultural dialog about how we work to define ourselves as part of a civil society.

2

PROGRAMMING THE CRISIS

The Police Drama in the Post-Network Era

On 26 September 1990, Steven Bochco's *Cop Rock* premiered on the ABC network. Bochco, of course, had already earned a reputation as a television innovator and genre-bending genius with series such as *Hill Street Blues*, *L.A. Law*, and *Doogie Howser, M.D.* But *Cop Rock* promised to exceed even these ground-breaking series and defy all generic logic: a police procedural-musical in which cops and criminals, judges and juries would occasionally drop the façade of gritty realism for song and dance numbers. The idea was met with a mixture of wary skepticism, incredulity, and fatalism on the part of critics, audiences, and industry insiders alike. As one owner of an ABC affiliate grumbled, "Everyone, even Steven Bochco, has to have a loser sometime" (Churchill 1990: 10). *Cop Rock* lasted only eight episodes and was finally cancelled in December 1990. Perhaps because of the series' oversized budget ($1.2–$1.4 million per episode), or its creator's unprecedented production deal with ABC (ten series over ten years), or simply the perceived audacity of its premise, *Cop Rock* has become the paragon of television's spectacular flame-outs, a cautionary tale about the dangers of unbridled excess.[1]

But the series was more than an ill-conceived misstep and its failure was more instructive than simply a moral lesson about moderation. Rather, the fact that *Cop Rock* was produced at all should encourage us to consider more closely the specific institutional and economic conditions that made it possible in the first place. In fact, experimentation within the police genre during the 1980s and 1990s was not simply a matter of artistic innovation by a handful of maverick writer-producers; instead, developments within the genre grew out of a larger reorganization of the commercial media industry, particularly as cable and satellite distribution technologies started to expand viewing options beyond the three networks (ABC, CBS, and NBC). In this increasingly crowded marketplace, the need for products that were *both* familiar *and* unique was felt ever more acutely. Generic familiarity would draw audiences in and novelty would (hopefully) keep them coming back.

In this context, a show like *Cop Rock* might be seen as a logical, even rational, product within a period of institutional ferment. More broadly, the fact that the police genre – one of the oldest and most recognizable genres on television – grew in popular appeal and was fertile ground for stylistic experimentation during a period of rapid institutional transformation, even to the point of wild failure on occasion, opens up important questions about how genres function in the shifting strategies of production and programming. What opportunities were opened up for stylistic experimentation, and why was experimentation perceived as necessary in the first place? What made producing a police drama an attractive venture in this increasingly unstable marketplace? What programming niches did these series fill?

Perhaps the most basic question that arises when considering the industrial components of the police genre is, quite simply: Why the police drama? Why did this particular form become so popular during this particular historical moment? There are, of course, a broad range of social and political factors that go a long way toward explaining the rise in popularity of police dramas in the 1980s and 1990s, and much of the rest of this study will focus on the cultural functions of the police genre as a popular public forum about crime, community, and citizenship. But before moving ahead with more explicitly discursive and textual concerns, it is necessary to first locate what Herman Gray (1995) has called the "enabling conditions" that have encouraged the proliferation of the police drama since 1980. The primary determining forces in this particular era of television programming and production are technological (cable, satellites, and VCRs), economic (diversification, conglomeration, etc.), regulatory (battles over technology and ownership), and discursive (new ways of defining the audience for programming). This chapter first considers the related technological, regulatory, and organizational developments that exerted pressure on the commercial television industry in the 1980s and 1990s and paved the way for what is now widely referred to as the *post*-network or *neo*-network era.[2] I will then consider more closely the police genre within a range of programming strategies developed by broadcasters and cable networks alike, demonstrating that the genre is a far more flexible and discursively open form than has previously been recognized by critics. In the end, the example of the police drama should serve as a basis for reconsidering the flexibility of all genres in the context of the post-network era.

A Whole New World?: Cable, Satellite, Broadcasters, and Regulators

The story of commercial television at the end of the twentieth century is often framed as one of crisis: the major broadcast networks struggling to remain vital and solvent in the face of increasing competition. The crisis that commercial broadcasters faced in the 1980s and 1990s was less a matter of failing to compete with new technologies than of trying to figure out how to manage and navigate changes that they knew were inevitable – and potentially quite profitable. While broadcasters (independent stations and network affiliates in particular) were indeed quite concerned with the potential for cable systems to cut into their sources of revenue, and urged the Federal Communications Commission (FCC) to monitor and regulate the growth and operation of cable systems in regional markets, the networks themselves were well aware of the institutional restructuring and programming changes that were looming, especially as cable continued to grow.

In this section I want to focus on the emergence of cable in the 1970s and 1980s, and the regulatory battles that ensued, as a way of illustrating the complex, often ambiguous, relationship between broadcast networks, network affiliates, and their cable "rivals." The reason for this foray into technology and regulation is to establish the ways in which the networks had *anticipated* the changes within this period at least a decade ahead of time, but had been regulated out of a position to drive those changes. Thus, the variety that we find in the police genre (and the sometimes wild experimentation) is less a matter of frantic or chaotic *reaction* to a situation out of the networks' control, than of active *participation* in a television landscape that had simply become more contested, complex, and diversified.

The narrative of cable's growth has often pitted the networks and their affiliated local broadcasters against cable providers. The reason for this acrimony stems from the threat that these non-broadcast entrepreneurs represented to the networks' (and regulators') plans to guide the growth and expansion of television toward their own interests. Whether it was the increased channel capacity of coaxial cable, or the allocation of additional broadcast frequencies via the UHF band of the spectrum, the television industry was capable of at least limited expansion from the beginning, and it was only the competing institutional and governmental visions for television's future that encouraged or deterred development of these additional potentials.[3] The networks had always been interested in expansion (via cable, UHF, or subscription television), and according to media historian Erik Barnouw, as the cable television industry grew in the 1960s and 1970s, "(t)here was growing awareness that a nationwide linking of cable systems could turn into a coast-to-coast subscription television system. Broadcasters were hedging their bets by investing in cable systems" (1982: 353). According to Maurine Christopher writing in *Advertising Age* in 1981: "CBS was moving full speed ahead as a cable MSO (multiple systems operator) in the late 1960s until the FCC put this territory, as well as domestic TV syndication, off limits for the big three networks. This crackdown came during a concentrated effort to curb the power of ABC, CBS, and NBC" (1981: 48). Thus, we can begin to see that the tensions between cable operators and networks was, at least in part, a product of government concerns about network oligopoly and an uncertainty about cable's future structure and purpose.

This tense relationship took on a slightly different hue in the deregulatory market of the 1980s. Under the free-market policies of the Reagan administration, the networks were ready to make good on their once-thwarted ambitions. In 1980, CBS, Inc., a four billion dollar media conglomerate (Slater 1988: 297), owned not only the most successful broadcast network, but also "five over-the-air stations, and 14 radio stations" and was still interested in the possibility of cable ownership recently opened up by the deregulatory attitude of the FCC under Mark Fowler (Christopher, 1981c: 48). In a similar way to CBS, ABC was making plans for diversification through both cable and Low-Power TV. As *Broadcasting* reported in February of 1981, ABC was "the first of the three big networks to join the low-power rush at the FCC" which had already received over 1,500 applications ("ABC Goes After Five": 60).

> The proposed low-power stations, however, are only one step by ABC into new television services. ABC already has an upcoming "Alpha" cultural service for cable in partnership with Warner Amex Cable Satellite Entertainment Corp. And this week, it announced plans for "Beta" a cable service with Hearst Corp. geared toward women.
>
> *("ABC Goes After Five": 60)*

At roughly the same time, CBS also tried to get into the cable programming business with CBS Cable, a precursor to Bravo, focusing on "cultural programming" such as ballet and theatre. But CBS Cable lasted only a year and cost CBS millions of dollars (MacDonald 1990: 269). Similarly, NBC got into the cable programming game with its entertainment channel, which met with an equally disappointing fate. Despite these early failures, it is clear that the major broadcast networks were determined to capitalize on the potentials that cable network ownership made possible through new distribution technologies and a loosened regulatory structure.

Of course, in some ways, these new technologies *were* challenging the terms of network hegemony, and the crisis at the networks has been understood in many of the same terms as the crisis that faced the major movie studios in the 1940s and 1950s leading up to and following the Paramount Decree of 1948. Just as the dwindling audiences and profits in Hollywood were understood as the product of government intervention (via anti-trust litigation) and television's assault on national leisure habits, the decline of the three major networks has been blamed on the incursion of these new technologies that diluted the television landscape irreparably. As John Caldwell has pointed out:

> By the late 1980s front-page stories in the national press were loudly trumpeting the demise of the networks, who were "under attack" – besieged by

> an array of new video delivery technologies. By the early 1990s, the networks were publicly wringing their hands, as victims of cable, of unfair regulatory policies, and of syndication rules.
>
> *(1995: 11)*

The introduction of satellite distribution systems provided independent broadcast stations and cable systems with the tools they needed to be truly competitive with the networks. Not only could producers of syndicated programming now deliver their product to independent stations within the span of one day, greatly increasing the relevance and current appeal of their programming (something only the networks could supply up to this point), but cable systems could now expand their packages exponentially. As Erik Barnouw points out, satellite distributed services, such as HBO, WTBS, A&E, CNN and, of course, MTV, "enabled an existing cable service to offer, under various arrangements, not the up-to-a-dozen program choices available from early cable television but scores of choices" (1982: 494).

How did the networks respond to these developments? Many critics contend that they turned a blind eye to cable and satellite technologies. For example, with regard to the rise of satellite transmission and the specter of pay-television, J. Fred MacDonald argues that the networks simply failed to comprehend the severity of the threat that HBO foreshadowed in 1975, smugly going about their business-as-usual:

> Network television seems not to have fully comprehended the threat inherent in cable. When Home Box Office in 1975 requested from the FCC the right to bounce its signal off the orbiting Satcom I satellite, the commission announced a public hearing at which all dissenting parties would have the right to protest before a final decision was made. If ever there was a time to bring out the top executives and lobby, this was it. Instead the public hearing produced *no network dissent.* To compound the error, the satellite leased by HBO was owned by the Radio Corporation of America, parent company of NBC.
>
> *(1990: 244, emphasis mine)*

By way of explaining this reticence, George Mair, author of *Inside HBO*, argues that the success of HBO was due in large part to the "stupidity of its competitors. This mute response to the FCC's invitation to protest HBO's request to go up on the satellite is a classic illustration of that stupidity" (1988: 25). Mair goes on to suggest that movie studios, theater owners, and the broadcast networks *as a group* "seemed not to understand anything about the new technology (pay television) and its implications" or simply ignored new players like HBO "because they apparently did not think that Home Box Office could make it financially" (1988: 25).

On one hand, it is easy to accept the possibility that the studios, theaters, and networks were willing to let HBO and its parent company, Time, Inc., undertake a major financial risk – to venture forth into mostly uncharted waters – without any real intervention (Wasko 1995: 76). But to imagine that the networks (not to mention the major movie studios and theater owners) were simply blind to the advances taking place within their own backyard, and taken by surprise by cable and satellite (or, worse, totally ignorant of them), is to underestimate the flexibility of the television networks as players in the larger entertainment industry.[4] In fact, the networks' proven interest in establishing their own cable programming outlets was a clear path toward participation in cable, despite the FCC's reluctance to allow them actually to own and operate cable franchises. As the 1980s got under way, and the FCC began to loosen the reins on ownership restrictions, the big three networks, far from being blindly anti-cable, were actually quite active in their search for competitive leverage in a new media marketplace.

"Cop Shows are the Old TV": Programming Police Dramas in the Post-Network Era

Ownership of cable networks (if not cable systems themselves) was one approach the networks took toward securing their survival in the changing television landscape. But how did these changes affect programming itself? In the same way that the networks anticipated the changes that were to define the structure of the industry (even if they weren't able to control the direction of those changes), their programming strategies also anticipated the changing realities of the media marketplace: in particular, the need for a diverse array of carefully marketed programs. In this section, I will argue that this need not only be a diversity of *types* of programs but also could (and did) play out within individual genres as well, such as the ever-present police dramas, legal dramas, and medical dramas that continue to dominate the airwaves. By looking at the police drama specifically as part of the larger programming strategies of the broadcast networks I will demonstrate that both the resurgence of the genre itself, and the newfound *variety* of approaches taken within the genre, suggest that the networks and the producers that supply them with programming were in a period of transition: aware of the new competitive forces driving the media marketplace while continuing to rely on tried and true generic forms for their products. More generally, this transitional phase for the networks highlights a handful of central tensions in popular culture: the interplay between novelty and familiarity, the balance between reaching specific or general audiences, and the tendency of popular culture to recycle itself in new forms and combinations.

These broader trends are illustrated quite well by the conception and production of *Hill Street Blues*. When Steven Bochco and Michael Kozoll met with NBC Programming chief Brandon Tartikoff, and Michael Zinberg, at the La Scala restaurant in Los Angeles in 1980 to discuss ideas for new shows, Bochco quickly dismissed Tartikoff's pitch for a new cop drama: "*Everybody* is sick of them. Cop shows are the old TV" (Tartikoff & Leerhsen 1992: 159). For Tartikoff and his boss at the time, Fred Silverman, the idea to do a cop show in the first place arose from the recognition that there were, at that moment, no cop shows among the top twenty programs which, according to Tartikoff, could have meant one of two things: "It could mean that no one wants to watch these kinds of shows anymore. Or it could mean that we've spotted an opportunity, a need, and that people are ready to take a fresh look at the genre – if the show is done right" (Tartikoff & Leerhsen 1992: 159). To do the show "right," the already reluctant Bochco and Kozoll insisted that they would need complete creative control and freedom from the meddling of the folks at network Standards and Practices. According to Bochco, "there was simply no reason to do another cop show, or another 'this' show or another 'that' show, the same old way we'd been doing every other goddamned show, and getting our brains beat out, 'cause people weren't watching" (Longworth 2000: 202). Tartikoff agreed to the producers' demands and *Hill Street Blues* emerged as one of the most daring and challenging series in the history of network television.

The story of *Hill Street Blues*' creation and eventual success has been widely circulated (Gitlin 2000; Stempel 1992; Thompson 1996) and has become something of a network television legend – a benchmark by which to measure the patience and good faith of network programmers. Part of this legend, of course, is the oft-cited fact that *Hill Street Blues* was potentially the lowest rated series ever to be renewed for the next season by its network.[5] By most accounts, there were two reasons for this unprecedented renewal: last-place NBC had little to lose in taking a chance on a provocative and challenging series, and the show received unequivocal critical acclaim (including 21 Emmy nominations). But perhaps the two most important factors had to do with the way that *Hill Street Blues* addressed the network's programming needs in this period of transition.

The first of these reasons is the looming reality (and promise) of cable. According to Brandon Tartikoff:

> *Hill Street* was getting a higher rating in homes that had pay cable than in homes that didn't. That meant that the people who had the most options, the people who could choose from what was then about a 20-channel cable universe, were seeking out our show to watch. To me that meant *Hill Street* was the kind of show networks would need to compete in the future, when cable would mean access to 50, 100, or even 150 channels.
>
> *(1992: 164)*

Tartikoff's emphasis on competition in the future indicates that, even if the networks hadn't been actively seeking ownership in cable enterprises, the connection between programming and new institutional realities was fairly clear to them by 1981. The second of these reasons has to do with the *quality* of the audience for

Hill Street Blues. Those people of whom Tartikoff spoke – those "who had the most options" – were typically younger and more affluent than the majority of viewers; they were the part of the population most interested in new technology *and* able to pay for the expanded choices in programming that it offered. Though the overall numbers for *Hill Street Blues* were low, the *demographics* were "disproportionately high," composed of "relatively younger, more male, more prosperous viewers whose attention was worth proportionately more to advertisers" (Gitlin 2000: 305). In fact, by the time of the network up-fronts (the annual industry gathering at which network schedules are revealed, and advertisers buy time on particular programs), NBC had sold all of the available ad space for *Hill Street Blues*, which had garnered more revenue than its low ratings would have otherwise justified. In the 1984–85 season, though *Hill Street*'s average weekly rating was 13% lower than CBS's *Knots Landing*, both series were able to command $200,000 for a thirty-second commercial spot. Furthermore, at a time when network viewing on the whole was declining (down 4 percent in 1984–85), NBC, on the strength of series such as *Hill Street* and *Miami Vice* was up 10 percent in the 18–49 demographic (Corliss 1985: 66).

Network programmers, of course, follow – and try to satisfy – the desires of advertisers to reach a significant share of the audience; they do this by scheduling programs that will attract the desired audience and serve as an appropriate environment for the advertisers' messages. The increasing presence of cable, satellite, and recording technologies in the home encouraged the entire television industry to reconsider its basic assumptions about how to reach audiences and how to structure the revenue stream that feeds the industry at the most basic level (i.e. how to establish price guidelines in the face of increasingly unstable viewing practices). Of particular interest to many advertisers, of course, were the younger and more affluent viewers for whom network television had never held much interest, but for whom cable and satellite represented a potentially attractive alternative. As Joseph Turow points out with regard to advertisers: "New cable channels allegedly viewed in the wealthier neighborhoods of the U.S. were encouraging general discussions about the future of media. How should the advertising industry approach the new media, and the people who use them, and why?" (1997: 39). For my own purposes, I suggest expanding this question and asking: In an effort to serve these advertisers, how did producers and programmers approach traditional genres, and the people who use them, within the framework of the new media?

Erik Barnouw has suggested that one response by the networks to their increasingly tenuous programming leverage was to recycle "long-trusted genres – sit-coms, often zany, increasingly obsessed with sex; and dramas of the tracking, subduing, and killing of enemies foreign and domestic" (1982: 513). Thus, Barnouw offers one way to understand the persistence of traditional generic forms (such as the police drama) on U.S. television during a time of institutional ferment: as a fall-back position, a proven quantity to which programmers in crisis can continually return to guarantee at least a minimal amount of success. This explanation suggests that there would be a great deal of similarity among different series within these genres – a path of least resistance that most series would follow. But Barnouw's position does not necessarily fit with the variety found within and between generic texts during this period. In other words, it does not help to explain the simultaneous existence and popularity of series as diverse as *Hill Street Blues* and *T.J. Hooker*, *Miami Vice* and *In The Heat of The Night*, *Cagney & Lacey* and *Hunter*, all of which could be placed within the broader category of "police drama," but which also have significant formal and ideological differences. And, as the earlier example of *Hill Street Blues* demonstrates, the historical fact of the genre's previous popularity and success does not guarantee that programmers and producers will be looking to replicate the formulas of the past. In fact, the opposite may be true. While success *does* breed imitation, it also breeds variety.

In order to begin to come to terms with the complex institutional role of the police drama in the 1980s and 1990s (among other genres), we must carefully consider the delicate and shifting balance between recognizability and distinction at work within these texts and how generic identity was used to anchor these series as a set of identifiable texts while simultaneously allowing for experimentation within the form. We should also consider the role of fragmentation itself on programming: in particular, how have changes in the way the audience is conceptualized (by programmers and advertisers) affected programming and production decisions? In other words, how do programmers and producers attend to different segments of the audience within the same genre? In order to begin answering these questions, I will focus on three broad but interconnected areas: stylistic approaches to the genre, programming for specific audiences, and recycling series through syndication.

Style: Cutting Through the Clutter

Style, of course, has been a central part of the police genre since at least as early as *Dragnet* – in particular, the need for "realism." Jack Webb's insistence on procedural realism – on the attention to even the smallest detail – is legendary.[6] In contrast, Joseph Wambaugh's *Police Story* strove for a different kind of realism in the 1970s. Through several novels and then the *Police Story* anthology series, Wambaugh focused less on procedure and more on the police as fully realized human beings with personal flaws and problems: the ordinary human in extraordinary circumstances. The job of policing – of catching criminals – was still a central narrative concern, but it was the personal pressures that impinged on the detectives' and officers' abilities to do the job that was the real dramatic core of the series. Similarly, in the 1980s, *Hill Street Blues* also presented us with flawed human characters, but used an ensemble cast to tell several intertwined stories. To complement the chaotic qualities of the plots, the series also employed an array of visual and aural codes to signify verisimilitude, most noticeably during the weekly roll-call sequence which opened each episode with an unsteady, handheld camera swinging from character to character, losing focus at times, and a thickly layered, cacophonous soundtrack.[7] These stylistic nods toward verisimilitude have been echoed in nearly every police series since the 1980s, even to the point of the extreme self-consciousness of *NYPD Blue*'s compulsively shifty camera or *Homicide*'s hyperactive jump-cuts. The varied uses of these codes of realism across series help illustrate how even the most basic and widely shared stylistic reference point of a genre is also a site of differentiation – a way to cut through the clutter of programming.

Over the last twenty-five years, of course, the emergence of hundreds of cable and satellite networks (not to mention the competition from feature films on home video) has made this clutter even more pronounced. To address the need for distinctive programming, one strategy of the networks was a move toward what John Caldwell has called "televisuality." For Caldwell, televisuality "emerged from a number of interrelated tendencies and changes: in the industry's mode of production, in programming practice, in the audience and its expectations, and in the economic crisis in network television" (1995: 5). Each of these factors relates back to the emergence of alternate distribution outlets (cable, satellites, and VCRs) in the 1970s. At the center of Caldwell's argument is the idea that the nexus of increased competition and emerging digital technologies encouraged and facilitated an "extreme self-consciousness of style" (1995: 4):

> With increasing frequency, style became the subject, the signified, if you will, of television. In fact, this self-consciousness of style became so great that it can more accurately be described as an activity – as a performance of style – rather than as a particular look.
>
> *(1995: 5)*

This "performance" of style took many guises: aspirations to cinematic distinction through both technique and the presence of an *auteur* (Barry Levinson, Michael Mann, David Lynch, Oliver Stone); through videographic excess facilitated by digital editing packages and influenced by such cable phenomena as MTV; through event programming such as mini-series; and through the "dangerous" exhibitionist spectacles of trash and tabloid programming.

Many of the most notable police series since the 1980s have relied on stylistic flourishes of this sort in order to set themselves apart from the increasing clutter of programming available via cable, satellite, or home video technology. *Homicide*, *Twin Peaks*, and even *CSI* emphasized their cinematic pedigree (through the presence of Barry Levinson, David Lynch, and super-producer Jerry Bruckheimer, respectively); *The Wire* built on the prestige of *Homicide* and added the reputations of its literary-world writing staff (Richard Price, Dennis Lehane, George Pelecanos); *NYPD Blue* and *The Shield* relied a great deal on salty language and partial nudity to call attention to themselves; *Cagney and Lacey* strove for event programming, focusing on exploitable current issues such as abortion and domestic abuse; *Miami Vice* and *Hunter* relied a great deal on aestheticized violence heightened through slow-motion sequences. *Miami Vice* further enhanced these sequences by adding a contemporary rock soundtrack. For many network offerings during this period – series such as *Miami Vice*, *Cop Rock* or *CSI* for instance – "style was no longer a bracketed flourish, but was the text of the show" (Caldwell 1995: 6). In these cases, the procedural realism of the police genre often took a back seat to the moments of visual, aural, and dramatic "excess" – the stylized, music video pulse of *Miami Vice* or the surprising eruptions of song and dance in *Cop Rock*.

Generic Hybridity: Categories That Matter

As the above examples might suggest, one of the keys to understanding the style of police dramas in the 1980s and 1990s is the increasing reliance on generic *hybridity*: the mixing of two or more generic types in order to create a seemingly new, differentiated text. Genres, as several critics have argued recently (Altman 1999; Neal 2000; Mittell 2001), are not only textual phenomena; they are complex discursive constructions formed by the intersecting activities of producers, marketers, audiences, and critics. Thus, the strategic mixing of genres can be investigated by tracing the industrial and critical uses to which generic categories have been put. For instance, *Hill Street Blues* has been labeled by numerous critics as a mixture of the police drama and the soap opera (due to its reliance on character-driven melodrama and serial narrative), though the producers and the network responsible for its existence also saw the series as a mixture of police drama and situation comedy – "*Barney Miller* outdoors," according to Brandon Tartikoff (1992: 158). The same can be said for *NYPD Blue* and *Homicide*, both of which rely on a complex blend of melodrama and humor within a mostly serialized structure. But the plots of these series all flow rather differently than traditional soap operas (daytime or prime-time), which rely on infinite deferral of closure in order to keep the regular viewer "hanging." These police dramas, on the other hand, combine the serial structure of their melodramatic plots with a more contained, case-based narrative, typically closing down these procedural arcs within one or two episodes.

More in line with this soap opera structure is *Twin Peaks*, which also relied on a mixture of melodrama and humor, but opted for a deferred narrative structure. This relationship to the soap opera, however, was mostly downplayed by the network in an effort to avoid the stigma often attached to the melodramatic form. As Marc Dolan demonstrates, the initial marketing of *Twin Peaks* "conditioned viewers to classify *Twin Peaks* as a detective story rather than a soap opera weeks before the series came on the air" (1995: 37). The widely recognized tagline for the series – "Who killed Laura Palmer?" – is pure detective-story fare and functioned to guide the interpretation of the series in that direction. For Dolan, this marketing strategy goes a long way toward explaining both the rather quick demise of the series (only two seasons) and the cultural capital of categories:

> One wonders what might have happened if *Twin Peaks* had been marketed as more soap opera than detective story. On the one hand, viewers might not have grown so impatient with the second season, but, on the other hand, they might not have watched the series in the first place, given the aesthetic snobbism that often greets the continuous-serial form.
>
> *(1995: 37)*

Interestingly, Dolan leaves out the possibility that there were other generic identities through which to interpret the series, such as the gothic horror story or the occult thriller, both of which would make sense given the auteurist identity of the series' co-creator, David Lynch.

In fact, the notion of hybridity should take into account more than just the textual features that may or may not be present in any given text. Rather, hybridity can also function as a matter of brand identity. In the case of *Twin Peaks*, David Lynch surely acted as a kind of brand name – a known commodity – to which viewers could be attracted. Not only could audiences come to the series as fans of detective stories or prime-time soaps, they could also come as fans of David Lynch's films (with the attendant textual categories of gothic horror and occult thriller attached as well). The producer-as-generic-marker has become an increasingly important element in the marketing of television series since 1980. As John Caldwell states:

> Part of the emergence of the quality myth in 1980s television was that television was no longer simply anonymous as many theorists had suggested. Names of producers and directors assumed an ever more important role in popular discourses about television. While Aaron Spelling and Norman Lear were already household names, other producer-creators like Michael Mann and Stephen Bochco began to be discussed alongside their actors and series in magazines and popular newspapers.
>
> *(1995: 14)*

An important addition to this realm of authorial distinction in the 1980s and 1990s was the increasing presence of feature film directors whose presence acts as a mark of distinction for the marketing gurus at the networks. "Even if for a fleeting season, this imported class and visionary flash promises to work wonders for network programming – at least when hyped in the right way" (Caldwell 1995: 17). Thus, the element of generic hybridity, conceptualized as more than just the mixing of textual features (e.g. the police series *and* the musical) recognized after the fact, must be understood as part of the larger marketing strategies of studios and networks, which includes the intertextual relationships that big-name producers, directors, stars, and writers bring to the table.

For the networks in the 1980s and 1990s, this kind of brand-name strategy was used increasingly to attract an elusive or underserved audience segment: namely younger, urban, affluent viewers who had more entertainment options and were usually less inclined to see television viewing as a worthwhile use of their time. In the case of *Twin Peaks*, the series was a clear effort on ABC's part to go after the 18–49 urban audience that is so valuable to advertisers. And the strategy seemed to work, at least for a while. In fact, the event status of the series' two-hour pilot (on 8 April 1990), directed by Lynch himself, was enough to earn a 21.7 rating/33 share and a fifth-place finish for the week (Jones 1990: 3D). Following that promising performance, however, the series dropped significantly in the ratings: by 19 April, its numbers had dropped from a 16.2 rating/27 share for the previous week (the series' second episode) to a 13.1 rating/21 share. ABC, the series' network, maintained however that "most of the dropoff was in viewers age 50 and above, but that viewers 18–35 still like *Peaks*" (Johnson, 1990: 3D). Clearly, while the size of the audience is a significant concern for advertisers and networks, the fact that they were losing viewers mostly over 50 years of age seemed to assuage their anxiety about the ability of their show to attract the right kinds of viewers.

Case Study – Miami Vice: Derivative But Fresh

Perhaps no other series illustrates this interplay of generic hybridity and the premium on younger, quality viewers better than *Miami Vice* – a series that was conceived as the merger of two distinct textual types: the cop-show and the music video. The apocryphal story of the series' conception illustrates the elegance of the concept for the series through a simple pair of words purportedly written on a note pad by Brandon Tartikoff: "MTV Cops." Taken at its most basic level, Tartikoff's memo neatly ties these two seemingly unrelated groups of text together to form a new combination. But it also opens up the issue of the network's felt need for innovation in the face of audience erosion. By the time *Miami Vice* was conceived, the networks had begun to take note of a hard-to-reach but significant portion of the audience that was migrating to the more specialized, irreverent fare of cable: young male viewers. As Harry F. Waters indicated in 1985:

> Much of the audience erosion experienced by the networks over the past decade can be traced to younger viewers who, conditioned by the avant-garde verve of such music video channels as MTV, find the network product hopelessly stodgy. Thus, the fusion of MTV production techniques with high-action cop shows, the theory goes, could help lure many of these defectors back to the fold.
>
> *(1985: 67)*

MTV itself, of course, was a hybrid form of programming designed to attract a particular audience: a mixture of recorded music and advertising imagery, aimed precisely at younger consumers. As MTV's original parent company, Warner Amex Satellite Company, claimed to its potential automotive industry customers in a September 1982 advertisement in *Television/Radio Age*: "MTV viewers are young drivers *and* buyers 18–34. They grew up with television. They grew up with music. So we put 'em both together to bring them MTV" (MTV 1982a: 5). According to a November 1982 advertisement in the same trade publication, this particular audience segment is especially difficult to attract with traditional (i.e. network) programming: "MTV's audience is among the elusive 18–34's who are always difficult to cover with traditional television and are now becoming harder to reach as more and more of them drop out of the network audience" (MTV 1982b: 13).

Within only a few years, MTV was able to deliver on their promises to advertisers. As E. Ann Kaplan recounts in her book, *Rocking Around the Clock: Music Television, Postmodernism, and Consumer Culture*: "By the end of 1983 the channel had $20 million in ad revenue, and figures for 1984 show more than $1 million a week in ad revenue, with an audience of 18 to 22 million" (1987: 2). Most of these viewers were aged 12–34 years old. An additional sense of how successful MTV was at capturing the attention of these young viewers comes from the Nielson Television Index. According to Nielsen, the average minute ratings for all regularly scheduled prime-time programs in May, 1987 was a mere 6.0 for men aged 18–34, compared to an average of 12.9 for all television homes (Papazian 1989: 80); but within that same age group, 72 percent of those surveyed by the Simmons Market Research Bureau were said to have viewed MTV during an average week (Papazian 1989: 51). This number was up from 57 percent as reported in the November 1982 advertisement in *Television/Radio Age*. Of course, these increased numbers could be attributed to an increase in the number of households receiving cable between 1982 and 1987. But seen as *percentages* of all potential viewers, regardless of cable penetration, it seems clear that MTV's popularity within the male 18–34 year-old demographic was experiencing significant growth. And with demographic indicators like these for such an elusive segment of the audience, it seems only reasonable that the traditional broadcast networks would try to find their way to these viewers via similar programming strategies.

In MTV, the producers of *Miami Vice* found the value of "mood." Relying on a mixture of cinematic, highly stylized images often evocative of *film noir* (again, hybridity is a central issue), a contemporary rock soundtrack, and a *mise-en-scène* that consisted of expensive Italian suits, Ferraris, and pastel-colored stucco mansions ("no earth tones" was one of the series' guiding visual principles), the series demonstrated that the police procedural need not founder in the detached, no-frills drudgery of *Dragnet* or in its opposite: the often inexplicable cartoon violence of action series like *Starsky and Hutch*. As Emily Benedek described the series in *Rolling Stone*:

> The design scheme is a juxtaposition of flashy high-tech (cars, guns, chrome interiors) with pastel colors and art deco lines of the restored South Beach area of Miami. In one scene from the pilot episode, following a long shot of Crockett and Tubbs in the Ferrari, the car rolls to a stop under an arching pink and blue neon sign that reads "Bernay's Café." Beneath the sign is a lone, lit telephone booth. Everything else is blacked out. Sonny gets out of the car and steps to the phone. Edward Hopper in Miami.
>
> *(1985: 56)*

Similarly, Lee Katzin, who directed the first season episode, "Cool Runnin,'" explains that the show "is written for an MTV audience, which is more interested in images, emotions and energy than plot and character and words" (Zoglin 1985: 61). Katzin's description echoes the thoughts of Robert Pittman, the founder of MTV, who once stated that MTV would "make you feel a certain way as opposed to you walking away with any particular knowledge" (Denisoff 1990: 241).

What is interesting to note, though, is that while the series did indeed rely a great deal on visual style to convey particular moods, the *music* of *Miami Vice* was actually used to enhance the narrative, rather than deflect it. While the series' creator, Anthony Yerkovich, admits that the show has often relied on extensive montage sequences set to music, shot much like music videos, he maintains that these sequences "are not gratuitous or extraneous to the story line but are designed to contribute to the dramatic narrative" (Smith 1985: C20). A typical example comes from another first-season episode, "Smuggler's Blues," which used the Glenn Frey song of the same name at several points within the narrative in order to highlight the action that viewers were seeing and to add another layer of meaning. As Fred Lyle, the series' music coordinator explains: "You're in a scene and all of a sudden up comes Frey, singing, 'It's the lure of easy money. It's got a very strong appeal.' And you're out again. It's like a Greek chorus coming in to chant: 'Fear him. Fear him!'" (Benedek 1985: 62).

The "Smuggler's Blues" episode raises another important connection between the series and MTV: cross-promotion and event programming. Not only did the producers use rock music in order to fill out the narrative and provide a contemporary feel for the series, but they also made frequent use of the musicians themselves as actors. A short list of musicians who appeared on *Miami Vice* includes Glenn Frey, Phil Collins, Ted Nugent, Little Richard, James Brown, and The Fat Boys (Brooks and Marsh 1999: 659). In the case of Glenn Frey, the series provided a unique opportunity for cross-promotion. Not only were both Frey and his song featured prominently in the episode (which shared its title with his song), but the video for the song played regularly on MTV and featured a number of montage sequences from the episode. Such a close connection between performer, song, and series, was actually quite rare however. For the most part these guest star appearances were simply special events for the series, opening up opportunities for new fans, including older viewers, to find the show by following a favorite performer to it.

By mixing the youth-oriented touch of MTV with a more traditional form of programming like the police drama in order to produce *Miami Vice*, NBC was not simply going after the elusive youth market. Instead, the strategy of the network seems to have been to *expand* its audience base for individual shows such as this one. In other words, the idea here is to *add* the younger viewers to the already existing, more stable body of network viewers without simultaneously turning those other viewers away. In fact, the goal all along was to generate big numbers for the series. As Sally Bedell Smith reported in January 1985 during the show's first season: "*Miami Vice* has attracted an average of only 22 percent of the audience this season – below the 26 percent considered the minimum for continuation. Nevertheless, the series has acquired a following among young viewers" (1985: C20). And NBC vice president for series programs, Jeff Sagansky, told Smith: "So far, the audience isn't big enough to make a hit, but once the drums get going, we think it will be a hit" (1985: C20). While it is tempting to think of new programming strategies at the networks as attempts to capitalize on audience fragmentation, it is necessary to keep in mind that, even in the era of 500 channels, programmers still need to balance both the quality and quantity of the audience.

In this sense, then, *Miami Vice* had to walk a tightrope between being both cutting edge (distinctive and appealing to its youth audience) and recognizably consistent with more traditional network fare. The result was an innovative program whose innovation was a matter of recycling recognizable forms rather than creating something wholly original. Tom Shales reported in 1986 that Brandon Tartikoff (the programmer most responsible for *Miami Vice*) "doesn't invent new programs; those few he does create are stitched together from scraps of old ones, or scraps of hits from other media. And he admits it" (1986: 72). Tartikoff himself explained that the secret to a successful show like *Miami Vice* was to make it "'totally derivative, yet seem fresh.' I mean, I use the term 'fresh' rather than 'original.' *Miami Vice* was really Hill and Renko with a music video feel. That was where the idea began" (Shales 1986: 72).

This situation is not unlike that which, according to Rick Altman, drives Hollywood feature film production. Altman argues that while most critical accounts of Hollywood film production focus on the idea that "generic templates undergird Hollywood's profit-assuring, assembly-line production practices, careful inspection suggests that Hollywood prefers romantic genre-mixing to the classical idea of generic purity" (1999: 129). This preference can be explained by virtue of the fact that the more generic markers a text can contain, the more it can attract disparate segments of the overall audience base. "Hollywood's basic script development practice involves (a) attempts to combine the commercial qualities of previously successful films, and (b) the consequent practice not only of mixing genres but of thinking about films in terms of the multiplicity of genres whose dedicated audiences they can attract" (Altman 1999: 129). Thus, the hybrid quality of network offerings like *Miami Vice* should not be seen simply as an attempt to program for a younger audience alone. Instead, it should be seen as an effort to *include* these viewers into the existing audience base for the network in order to satisfy advertiser desires for economies of scale. Tartikoff's famous two-word note is not unlike a high-concept pitch to a movie studio – reducing the idea for a program to its most basic elements, a mixture of two already recognized and successful forms, each of which is capable of attracting a particular segment of the audience.

Diversity in the Audience

While *Miami Vice* was designed in part to entice younger male viewers back to the traditional networks, other police dramas were also designed to attract other segments as well: in particular older viewers and women. By way of example, I will consider two popular police dramas from the 1980s and 1990s: *Cagney and Lacey* and *In the Heat of the Night*.

As Julie D'Acci has shown, the conception and development of *Cagney and Lacey* stemmed initially from a desire to engage with the discourses of the liberal women's movement in the U.S. – a design that fitted within the larger desire of the television networks to provide relevant programming, especially for new working women (1994: 4). As D'Acci details brilliantly, while the initial impulses of programs like *Cagney and Lacey* "were in sharp contrast to conventional images of women" in the 1970s, the 1980s saw a significant and sustained backlash against feminist movements which influenced the directions that feminist (and feminine) representations on commercial television would take (1994: 4). While the initial conception of *Cagney and Lacey* was as a traditional cop show, employing "many conventions of the police genre, including police procedural, violent action, a New York setting, present-day time, and antagonistic commanding officers" (D'Acci 1994: 119), as the series progressed (and faced a number of significant challenges including cancellation), the series was forced to change its approach. Of particular importance here is the move that the series' creators made from cop-show to women's program – from the traditional police procedural to the exploitation programming more akin to the "Movie of the Week," again highlighting the importance of hybridity in the programming strategies of the networks.[8] As D'Acci points out:

> The series became a hybrid, a combination of police drama and more traditionally women-oriented forms. In the textual operations of this hybrid genre, women were produced in a number of different and often contradictory ways – as active gender-trangressive cop/heroes (although in modified

> form after the network's interference), as autonomous protagonists, as a father-identified single woman, as a nurturing mother, as victims (of rape, violence, sexual harassment, cancer, and alcoholism), as heterosexual love interests and as traditional female TV comics.
>
> *(1994: 133)*

The hybrid quality of *Cagney and Lacey* addressed the demands placed on network programmers by the increasing popularity of cable outlets such as Lifetime. In fact, the development of *Cagney and Lacey* preceded and anticipated the development of Lifetime in 1984, much as ABC's proposed Beta channel had just a few years earlier. As Jackie Byars and Eileen Meehan point out, several conditions led to the emergence of the "working woman" as a valuable commodity on the heels of the rise of second-wave feminism, among them the decision by the A.C. Nielsen Company in 1976 to include working women as an official category in its ratings methodology, and an economic recession that encouraged women to enter the workplace in order to provide a second income for their families (Byars & Meehan 2000: 150). "With upscale working women specifically measured by the ratings, advertisers could and did demand access to this attractive audience, and networks began to program for both upscale men *and* women" – a strategy that encouraged hybrid series such as *Cagney and Lacey* that combined the masculine police drama with the more female-centered exploitation melodrama (Byars and Meehan 2000: 151). As with *Miami Vice*, the programming strategy for both *Cagney and Lacey* and Lifetime was to *expand* their viewership to include both men and women. A similarly inclusive strategy was part of the idea behind Stephen Cannell's *Hunter*, which features a rare male/female partnership at the center of the narrative. This partnership allowed the series to combine the aestheticized violent police action of *Miami Vice* with exploitation narratives of rape and domestic abuse in the spirit of *Cagney and Lacey*.

While both *Miami Vice* and *Cagney and Lacey* were designed to capture elusive "quality" segments of the audience (young males and upscale women, respectively), another series, *In The Heat of the Night* was specifically designed to attract an audience that had long been neglected by network programmers seeking the urban youth market (and was being all but ignored by cable networks): specifically, viewers over 55 years of age. Though cable pioneers like MTV's Robert Pittman spoke about paying heed to diversity among consumers, especially in "the split between those who grew up with TV and those who didn't" (Taylor 1981: B8), most programmers continued to ignore that split in favor of the younger end of the spectrum – the TV generation – as the "key to cable success." *In The Heat of the Night* grew out of the opposite impulse.

Following his departure as head of NBC in the early 1980s, programming wizard Fred Silverman started a new life as an independent producer on a string of hit series, each of which featured an aging television star with wide appeal: Dick Van Dyke (*Diagnosis Murder*), William Conrad (*Jake and the Fat Man*),

Andy Griffith (*Matlock*), and Carroll O'Connor (*In The Heat of the Night*). These series presented a relatively risk-free approach to programming and production as their stars almost guaranteed a significant share of the audience. But that share consisted mostly of viewers over 55 about whom advertisers are typically less enthusiastic. But in a diversified television landscape, logic held that a network could successfully counter-program with a series that appealed to a mostly under-served portion of the audience. "We offer the only show with older appeal. It offers a whole new audience at 8 p.m. on Friday," said Silverman in a 1989 interview (Gubernick 1989: 85). Actually, *In The Heat of the Night* emerged at a moment when series with older demographics, such as *Murder, She Wrote* and *The Golden Girls*, were consistently finishing in the top ten programs each season. A series like *In The Heat of the Night*, then, must be seen not only within the counter-programming strategies of the networks, but also as part of the never ending search for untapped segments of the audience that might supply networks and advertisers with unexpected revenue streams.

In a 1992 interview Silverman maintained that a network "can't have 22 hours of *The Fresh Prince of Bel Air* or *Blossom*. You need some diversity. And there is an enormous amount of buying power in older Americans" (Giltenan 1992: 14). Surely, this sense of the buying power of older Americans initially encouraged the production and scheduling of Silverman's series. And, indeed, both *Matlock* and *In The Heat of the Night* were enormously popular among older viewers. In 1992, Nielsen reported that these two series finished, respectively, 40th and 28th in total households, and 5th and 3rd among viewers over 50. But they finished only 83rd and 65th, respectively, among the all important 18–49 demographic (Battaglio, 1989). That same season, both series were dropped from the NBC line-up.

The shift from embracing highly rated but older-skewing shows, to dropping them in favor of demographically younger shows, may be directly attributed to the growth of Fox as the fourth broadcasting network. The success of Fox was tied directly to its marketing and appealed to those younger viewers that shows like *Miami Vice* had been designed to capture. With successful series like *Married, With Children*, *The Simpsons*, and *21 Jump Street* – a police drama with nothing but new, young talent – Fox ate directly into a key segment of the audience that the other three networks had been working hard to attract. With those viewers on the move, the networks could no longer afford the luxury of resting on high household numbers if it meant sacrificing "quality" audiences. As Stephen Battaglio reported in *AdWeek* in 1992, Robert Niles, the senior vice-president of marketing and research at NBC at the time, stated that the network "is willing to risk household ratings if its primetime lineup can become more attractive for ad dollars. The network's priority for next fall will be shows that hit with adults 18–49 – the favorite demographic for advertisers" (1992: 4).

The point to be made here is that television programming, particularly as it was forced to meet the demands of advertisers in an expanded television landscape, is always balanced between two competing and ever present tendencies with regard to the audience. On the one hand, there is increased incentive to focus on particular segments of the audience in order to assure advertisers of an efficient use of their money; this we can call a matter of *exclusivity*. On the other hand, both traditional broadcast networks and, increasingly, cable networks are still interested in capturing as large a slice of the audience as possible – a matter of *inclusiveness*. This tension is captured nicely by Beth Barnes and Lynne Thompson:

> The logic of media specialization requires an emphasis on advertising vehicles that deliver more homogenous (and therefore smaller) audiences over vehicles that deliver larger (and therefore more heterogeneous) audiences. Cable audiences were more homogenous (and smaller), as documented here, but the advertising industry had yet to fully relinquish its emphasis on audience size.
>
> *(1994: 91)*

As early as 1984, cable networks like USA, CBN, and Ted Turner's Superstation, WTBS, were leaving their smaller, specifically targeted audiences behind in the search for broader, more demographically diverse viewers. As Susan Spillman reported in *Advertising Age*: "The goal is less to be innovative than to be watched" (1984: 87). One way to capture these viewers was by procuring off-network or syndicated programming.

Reduce, Reuse, Recycle

In light of Brandon Tartikoff's equation between being "derivative yet totally fresh," one of the surest methods for commercial television producers and programmers to guarantee success is through recycling television's own past. Television has almost always been – at least in part – a "repeat" medium, dating back to its days as a showcase for Hollywood films. Reruns, of course, have always been a significant reality in television programming. Indeed, entire networks – for example Nickelodeon and TV Land – exist to recycle key programs from television's past. Hybrid shows like *Miami Vice* recycle and recombine popular genres (police drama and music video), visual styles (*film noir*), and popular music. Likewise, *In The Heat of the Night* recycled a beloved performer (Carroll O'Connor) and the prestige of its antecedent text: the 1967 Best Picture Award-winning film of the same title starring Rod Steiger and Sidney Poitier.

And, of course, these police dramas do not exist only at the time of their original production: they are continually recycled across the television landscape – placed into new, sometimes seemingly incompatible contexts. In other words, they have a certain currency that outdistances their original network value. For example, though *In The Heat of the Night* was inauspiciously dropped from NBC's line-up in 1992 because it didn't connect well enough with younger audiences, that same year it was picked up for syndication in 146 markets representing 86 percent of stations in the U.S. ("MGM Domestic Television" 1992).

In fact, cable networks have long been interested in recycled material from the networks. This move by cable networks to broaden their horizons did not mean the death of narrowcasting logic. What it meant instead was a new understanding of narrowcasting – a way of expanding the idea to take advantage of existing programming through the carefully considered recycling of off-network or syndicated series in order to fill holes in their schedules. As the vice-president of programming at the USA network stated in 1984: "At this point, for the money, we can get better quality programming off the networks or from syndication than we can producing it ourselves" (Spillman 1984: 87). But while the cable networks were seeking off-network programming, they also worked very carefully to re-purpose and position that programming for their ideal audience. For example, in 2002 Bravo, a cable network devoted to the arts, stripped the entire run of *Hill Street Blues* under the banner: "The Art of Television" – clearly capitalizing on *Hill Street Blues*' long-standing reputation as one of the most innovative programs in television history. Similarly, TNT has programmed reruns of *NYPD Blue*, *In The Heat of the Night*, and *Law & Order* (among a range of other, non-police dramas) as a way of advertising themselves as the place for drama: "TNT – We Know Drama." Perhaps the most obvious example of re-purposing for a specific audience is Lifetime's brief encounter with NBC's *Homicide: Life On the Street.* As I have written elsewhere (Nichols-Pethick 2001) the network went to great lengths – cutting specific scenes and repositioning commercial breaks – in order to re-structure the narrative for a Lifetime audience. What seems clear is that, while these police series were created and programmed by the broadcast networks in order to address a specific range of programming needs, the form is still flexible enough to allow these series to travel across programming formats and retain their value for a new set of networks and viewers.

The issue of recycling does not stop at re-purposing old series for new contexts. Cable networks, beginning in the 1980s and 1990s, also began to create their own original programs as a way of reaching out to their intended audience. What is significant about these endeavors is the fact that they so frequently relied on traditional forms for their ideas. As an early example, in 1989, David Hall, then general manager at TNN (known at that time as The Nashville Network) discussed the cable network's plans for a new cop show, *Nashville Beat.*[9] While the series was designed to capitalize on the success of *Miami Vice*, it was also designed to meet the needs of TNN's audience which, according to Hall, is "conservative and family-oriented...The cop show fits our programming philosophy. It's not *Miami Vice*. The bad guy gets caught and always loses" (Battaglio 1992: 4). Interestingly, the series was set to star Kent McCord and Martin Milner, the two principle stars of *Adam-12*, Jack Webb's most successful cop show after *Dragnet*. The idea for such a series suggests that the commercial television landscape in the 1980s and 1990s was caught between two competing tendencies – to capitalize on the loyalty of a small but stable body of differentiated viewers on the one hand, and to program for the broadest possible audience on the other by recycling recognizable quantities from the past, both in terms of the stars and the genre itself.

Conclusion

As the above examples suggest, the police dramas of the 1980s and 1990s were far from a simple set of texts. In keeping with my central claim that the police genre is a far more varied and flexible discursive form than has previously been acknowledged, I have argued that it makes less sense to try and explain the persistence of the genre as a whole than to highlight the ways that the genre has been used in a number of disparate programming strategies. Generic texts do not simply function in one way at all times and in all situations. Instead, genre is a *tool* used by producers and programmers alike to anchor programming decisions – to provide some security and certainty while simultaneously trying something new in the ongoing effort to reach as large and as valuable an audience as possible. This has always been the dialectic of generic texts: the simultaneous pressure to be both recognizable and distinct. This emphasis on the ways in which the police genre was activated (and re-activated) across a range of programming strategies points to both the ordinariness of the form and to its openness.

Television production and programming are always balanced between the need for familiarity and novelty. This tension is not only evidenced at the level of generic types but in the very essence of television as experienced by audiences. Novelty exists most apparently in the flow of television – the movement from one program into another either via the schedule itself or as activated by the audience though channel-switching.[10] At the same time, the need for familiarity anchors the flow of programming in the comfortable movements of genre and formula. It is this dialectic between distinction and recognizability that is definitive of the television viewing experience and that drives programming decisions in the first place. And it is this dialectic that the present chapter has attempted to address in terms of the uses to which police dramas (among other generic forms) were put by producers and programmers during the tumultuous decades of the 1980s and 1990s and then into the 21st century.

The growth and expansion of the police drama during this period is symptomatic of a larger reorganization of the commercial television industry in the U.S. In other words, the variety and flexibility that I am arguing for in critical approaches to this genre can first be recognized in the programming strategies of the networks themselves during this period of rapid and often confusing ferment. While the "story" of the police genre (or almost any genre, for that matter) is typically constructed in terms that are both textual and evolutionary (the birth and maturation of *form*), my own approach engages with the contradictions, tensions, and anxieties that accompany any historical era.[11]

My goal in this chapter has been twofold. My first goal was to address the question of how the police genre served as a symptomatic illustration of the changes in production and programming strategies of commercial television during the 1980s and 1990s. But my second and more central goal was to demonstrate, through these strategies, that the police drama is far more varied and flexible than has previously been acknowledged in textual criticism of the genre. In order to accomplish these goals, this chapter began with a consideration of some of the larger changes in production and programming that helped shape programming strategies during this period: issues such as the battles over cable (and the regulatory issues that helped shape these battles in the first place) and the segmentation of the audience into smaller and smaller niches (with the accompanying changes in marketing programs to particular audiences). These changes helped shape production by simultaneously encouraging expensive, high-risk programming that emphasized genre-mixing and seriality, as well as more cost efficient, lower-risk programming with more recognizable stars and formulas. In the end, the fact that the police genre in the 1980s and 1990s was activated within a wide range of programming strategies, by both traditional broadcasters and cable networks, underscores the genre's potential *variety* and allows us to reconsider how generic categories can (and *should*) be considered in light of this flexibility and variety.

3

THE POLICE DRAMA IN TRANSITION

Reconstituting the Cultural Forum in the 1980s

In the pilot episode of *T.J. Hooker* the title character, played by William Shatner, sits in a bar surrounded by his young and eager recruits winding down after a hard day of academy training. Despite the general good mood of the gathering, Hooker takes an opportunity to tell the story of a young rookie killed during his first week on the streets and to wax nostalgic for days gone by: "You know what I missed at the kid's mass? Latin. I wanted to hear Latin instead of everything in English. What about doctors who made house calls, repairmen who knew how to fix things, ballplayers who hustled, boxers who would get carried out of the ring instead of quitting because they had stomach cramps?" Hooker's mood darkens suddenly; he looks directly into the camera and shifts emotional gears: "Or the death penalty for hoodlums who snuff a life out like it's so much garbage. I've seen the past, gentlemen, and it works." With that, Hooker exits, leaving his young charges to ponder their future as police officers and their present commitment to Hooker's vision of justice.

Hooker directs his diatribe at his immediate audience – the collection of rookie cops who lack his experience and wisdom. But his words are also calculated to strike a chord with the audience of television viewers who share his frustrations and nostalgic point of view. Hooker's speech has a clear narrative purpose in establishing his authority over the rookies, but it also responds to real social and political concerns in the larger cultural debates about what constitutes crime, community, and citizenship. His statement addresses the complexity of the world outside of television and, as such, "provide[s] a rhetorical response to actual social concerns" (Parry-Giles & Traudt 1991: 145). Hooker's statement goes unchallenged in the episode (a fact that signals the more conservative ideological tendencies of the series as a whole), but he is nevertheless participating in a larger dialogue that occurs in American society and, more specifically, in the police genre.

In keeping with my central argument that the cycle of police dramas since the 1980s constitutes a "cultural forum" or arena for a larger dialogue about crime and punishment,[1] I see Hooker's speech serving a dual purpose. First, it is a strident example of the conservative strain in police dramas, often a reaction to a legal system represented as increasingly liberal or "soft on crime," especially around issues of due process and rights of the accused.[2] Second, it demonstrates the way in which a television series like *T.J. Hooker* takes a position in an ongoing dialogue about criminal policy: a dialogue in which all police dramas participate. Hooker's perspective is a pointed reaction to a growing trend in popular representations of the police during the 1970s and 1980s, one that complicated our understanding of the social function of the police, the fairness of the legal system, the responsibilities of citizens, and the causes and consequences of crime.

As participants in a cultural forum, police dramas not only address ongoing social debates about crime, but also respond to developments within the popular genre of police fiction in television, movies, and literature. When Hooker invokes a more noble and reassuring past, he is also implicitly calling for a return to a more traditional form of the police genre: one that existed before revisionist storytellers gave us fallible, morally compromised cops whose authority has been diminished by unresponsive bureaucracies and a more liberal judiciary. In order to understand how a television series takes shape and finds an audience at a particular moment in time, we should take care to describe how it responds to the history and ongoing development of its genre and then how it represents particular positions in contemporary debates about criminal policy and social issues.

All popular texts respond in some way to ideas that are circulating in the culture at large in order to capitalize on the issues and beliefs that audiences are most likely to respond to. This is what John Corner refers to as "the extraordinary cultural dynamics" of popular cultural forms such as television – the simultaneously centripetal and centrifugal forces that allow producers to pull ideas from the "culture-at-large," filter them through various generic structures, and return them to the broader culture in the form of stories (1999: 5). Todd Gitlin captures this dynamic in his discussion of how network executives work to gauge the "cultural mood" by continuously trawling for ideas in other popular cultural channels such as films, bestsellers, daily newspapers, and magazines:

> The network antenna is always rotating because, in the end, all the testing and copying and recombining and inside-track planning in the world aren't conclusive. Nothing can dissolve the network's dependence on certification by a mass audience. The trick is not only to read the restless public mood, but somehow to anticipate it and figure out how to encapsulate it in a show. No one comes to such arcane work innocent of ideas about what the market will bear, ideas that circulate constantly through the standardized channels of executive culture. The executive "instinct," much praised in the

> industry, is a schooled instinct, formed in experience and concentrated by that common culture.
>
> *(2000: 203–204)*

Part of this "schooled instinct" is the reliance on generic forms that allow for efficient production of series texts and that correspond to the ideas circulating in "the ebb and flow of popular feelings" (Gitlin 2000: 205). Thus, if the popular mood includes concerns about law and order (and when does it not?), one could reasonably expect a turn toward these issues in popular culture. Since the police drama has remained a fairly regular staple on network television (and beyond) it offers a valuable opportunity to think critically about how it has responded to changes in the popular mood, under what conditions, and how these changes manifested themselves within a given period: in this case, the decade of the 1980s.[3]

Making a Difference in the 1980s

Because new police series respond to the history of the genre and to other forms of police fiction in the cultural marketplace, the ideas that drove the police dramas of the 1980s were, in part, responses to innovations occurring across a broad spectrum of popular cultural sites in the 1960s, 1970s, and early 1980s. For example:

- motion pictures such as *In the Heat of the Night* (1967), *Bullit* (1968), *Dirty Harry* (1971), *The French Connection* (1972), *Serpico* (1973), and *Fort Apache, The Bronx* (1980);
- novels such as Ed McBain's "87th Precinct" series beginning with *Cop Hater* (1956), and Joseph Wambaugh's novels about the LAPD beginning with *The New Centurions* (1971);
- other television programs: *The Streets of San Francisco* (1972–77), *Police Story* (1973–77), *Kojak* (1973–1978), *Police Woman* (1974–78), *Starsky and Hutch* (1975–79), *Baretta* (1975–78), and perhaps most surprisingly, the situation comedy *Barney Miller* (1975–82).

Even documentaries, such as Alan and Susan Raymond's *The Police Tapes* (1976), which aired on PBS, had a direct influence on the style and subject matter of *Hill Street Blues*.[4]

The influences of these texts on the police dramas of the 1980s are a far reaching and often contradictory combination of themes, visual styles, narrative devices, and character types. A partial list of developments would include: the decaying urban space as the location for crime; the sense that the police are the last, thin line of defense in the fight against that decay; a focus on the personal interior lives of the police (the way that the job interferes with and changes their goals, aspirations, and outlooks); the heroic loner working (sometimes violently, but always effectively) outside the system which is seen as increasingly flabby and incapable of dealing adequately with crime; a reliance on violent action (gun battles, car chases, etc) to close cases and provide resolution and justice; the debt to procedural realism in the depiction of the job; the political opportunism of the upper ranks in the department and the increasing knowledge gap between the cops dealing with the "reality" of the street and the politicians who bend crime policies to fit political agendas; the police station as a social microcosm comprised of a wide array of character types, races, religions, ethnicities, and ages, as well as the introduction of gender as a key variable in the person of the cop.[5] Obviously, no single series could contain everything included in this (still abbreviated) list. What it suggests is that police dramas, like all popular forms, are not bound by a strict and narrow set of generic rules but, rather, pick and choose from a wide range of possibilities and opportunities that help shape the look and feel of each series as well as individual episodes.[6]

In what follows I trace discursive possibilities and choices across a range of series that were popular in the 1980s. These possibilities and choices include areas of narrative structure, character development, and style of presentation; but they are products, first and foremost, of social discourses in transition.

Social Discourses in Transition

The police dramas of the 1980s were produced within the context of broad social and political upheavals that affected the work and theory of policing. The key developments that I will address here are the economic decline of major urban centers in the U.S., the implementation of *Miranda* warnings, the rise of the Victims' Rights movement, the increasing popularity of community policing initiatives, and the "war on drugs." These developments provide an important foundation for understanding the various narrative concerns taken up by the television police series of the 1980s and 1990s.[7]

Many representations of the police in the 1980s – in books, movies, the news, and television drama – were placed against the backdrop of widespread urban decay. Films like *Fort Apache, The Bronx* and *Escape From New York* (both produced in 1981), for example, depicted New York City as a nightmare landscape. While the latter of these two films is a science-fiction fantasy set in a dystopian future (in which the boroughs of New York City have been turned into a maximum security prison for society's most violent felons), *Fort Apache, The Bronx* showcased the South Bronx in its contemporary state in 1981: as the charred and abandoned skeleton of a city, made infamous by President Carter's "sobering" tour in October 1977 of Charlotte Street in perhaps the most devastated section of the borough, and Howard Cosell's famous announcement during the 1977 World Series that "the Bronx is burning."[8]

Fort Apache, The Bronx may have been influenced by Alan and Susan Raymond's 1977 television documentary, *The Police Tapes*, which featured the officers of the

44th Precinct in the South Bronx – a precinct beset by crushing poverty and rampant violent crime (fifty-one murders in fifty-one days).[9] Both films emphasize the sense of defeat and frustration that follow these cops around; they also highlight the devastated landscape within which the action of the police seems almost beside the point. In the final scene of *Fort Apache, The Bronx* for instance, as two officers, Murphy (Paul Newman) and Carolli (Ken Wahl) chase down a burglary suspect, the camera follows the action but occasionally pauses at the sight of a murdered body lying in a heap of rubble or a building being demolished by a crane. As the officers close in on the suspect, Murphy dives forward and the image freezes, suspending him in mid-air; only the sound of a voice – a scream (of delight? pain?) – carries the action forward. Importantly, we do not see the officers apprehend the victim: the film offers no moral reward at the end of the chase. Instead, the still frame simply fades to black and comes up again on a panning shot of the same demolished cityscape as the credits roll. Similarly, the final sequence of *The Police Tapes* shows an apartment house burning as firefighters work to douse the blaze. The images of the firefighters are eventually replaced, however, by a series of static images of the demolished remains of the building in the aftermath of the fire – police sirens wailing in the background, on their way (it seems) to something else. In both films, we see the heroic action of civil servants (police and firefighters) cut short: resisting any kind of tidy resolution. While both films want to portray the police heroically, the overriding message is one of defeat and frustration. As Murphy says to his commander, Connelly (Ed Asner): "I'm feeling as burned out as those damn buildings down on Charlotte Street."

That sense of defeat and frustration at the near impossibility of truly heroic action in an environment of rapid decay also permeated the world of *Hill Street Blues*. In an episode from the first season entitled "Gator Bait," Lt. Howard Hunter, the leader of Hill Street Station's Emergency Action Team (EAT) – a SWAT-like paramilitary operation – waits with members of his team in an underground sewer system. They have been sent down into the bowels of the city to search for and remove the alligators that are posing a threat to sanitation workers. Howard and his men are forced to wait for a representative from the SPCA (Society for the Prevention of Cruelty to Animals) to arrive with tranquilizer darts so that they may deal with the alligators in a humane fashion. The SPCA representative never arrives (he gets into a traffic accident and "darted" with his own tranquilizers) and Howard and his men are left to contemplate their increasingly ridiculous metaphoric situation. "If I were writing my memoirs, Nelkirk," says Howard, "I'd call this little episode 'Prelude to Oblivion.' It's a symptom of this city's deteriorating nervous system. What galls me is that there is still time enough to save it."

Of course, this scene – and much of the series – is intended as comedy: clearing the sewers of alligators is hardly the stuff of action dramas, and Howard's Emergency Action Team is clearly a parody of the paramilitary approach to policing that became so popular in precincts across the U.S. in the 1970s. But the

sentiment is still vitally important to the series. Just how, or even *if*, the city can be "saved" is one of the key themes of *Hill Street Blues*. The vision of the city painted by the series is grim and fearful. Located in no city in particular (but clearly emulating the South Bronx), the streets of *Hill Street Blues* are the rotting veins of America's urban industrial landscape of the 1970s: the logical outcome of an economy in peril. It is a metaphoric modern nightmare, replete with gang warfare, burned-out storefronts, deteriorating public utilities, and a civic service overseen by increasingly absent and self-centered bureaucrats. Random violence lurks around every corner and even bursts into the station house. The police themselves are left to react to the violence and try to keep the community from crumbling completely.

Like *The Police Tapes* and *Fort Apache, The Bronx*, the police officers on *Hill Street Blues* are not only the protectors of the city, but sometimes its victims as well. In the pilot episode, after Hill and Renko respond to a domestic violence call (discussed in the introductory chapter), they emerge from the building and find that their patrol car has been stolen. They search for a phone in order to call for a backup vehicle but the pay phones on the street have been vandalized and are inoperable. Renko angrily declares his disgust at the state of this neighborhood where not even police cars and pay phones are safe, and an ominous crowd begins to gather near the scene. Hill tries to keep his partner from making a bad situation worse, and leads him into the foyer of an abandoned tenement building. Inside, they encounter a small group of junkies, one of whom pulls a gun out and shoots the two officers, leaving them for dead. The scene is intended to shock viewers who have come to identify with Hill and Renko after their skillful handling of the domestic dispute, and the impact of the scene is heightened by the use of slow motion to show the officers twisting in agony as the bullets enter their bodies.[10] The shock of violence against the police is repeated in a later episode from the first season ("Up In Arms") when another officer, Harris, is murdered by a prostitute during an arrest. The same slow motion technique is used as the prostitute pulls a razor from her belt and slices Harris's neck when he moves toward her.[11] In both cases, the impact of the violence lies in the combination of its shock value and its randomness. The junkies and the prostitutes are not central characters in the episodes; they are merely part of the blighted backdrop, signs of the city's deterioration. Making the police the literal victims of this deterioration highlights the question of whether they can save the city from its sharp decline.

Other series approached the issue of urban decay from slightly different angles. Whereas the *mise-en-scène* of *Hill Street Blues* emphasized the squalor of the city streets, *Miami Vice* was intended to "evoke a Miami that mingles lavish living and mysterious underworld activities" (Smith 1985: C20). Much of the dramatic action in the series took place amidst the noteworthy architecture of the city, such as the Spear House and the Atlantic Condominiums, and featured wealthy drug lords (rather than junkies or street-level dealers) and undercover detectives, all perfectly suited to the environment. Michael Mann told one reporter: "When we

filmed in the Spear House it was no accident that one of our characters appeared in a pale turquoise shirt against the pink ochre walls" (Smith 1985: C20). While this careful production design was intended to give the series a specific aesthetic appeal unlike anything else in prime-time, it also played an important role in relocating the spectacle of decay *underneath* the clean, pastel surfaces of the city. This hidden decay – the underworld narcotics trade – more like a cancer than a bruise, threatens to destroy the city and its citizens from within. "Beneath appearances lie only more appearances...duplicity and vice are pervasive; they infuse and corrupt everything, including the Law" (R.L. Rutsky in Buxton 1990: 147).

On the surface, one of the key signifiers of urban decay in the 1980s was the increased level of drug use among the young and poor in America's inner cities; *Miami Vice* was largely concerned with what went on below the surface. In particular, the series largely ignored one of the central tenets of the "war on drugs" during the Reagan era – "Just Say No" – and focused instead on the drug trade as a "confrontation between the law and trans-border commodity flows" (Buxton 1990: 149).[12] According to series creator, Anthony Yerkovich, it was this international angle that first attracted him to the idea:

> I thought of [Miami] as a sort of modern-day American Casablanca. It seemed to be an interesting socio-economic tide pool: the incredible number of refugees from Central America and Cuba, the already extensive Cuban-American community, and on top of all of that, the drug trade. There's a fascinating amount of service industries that revolve around the drug trade: money laundering, bail bondsmen, attorneys who service drug smugglers. Miami has become a sort of Barbary Coast of free enterprise gone berserk.
>
> *(Zoglin 1985: 61)*

By taking this angle – that the drug trade is a diversified industry designed to sell a product – the series at least opens up the possibility of engaging with one of the larger causes of the drug "epidemic" threatening the nation at the time: the fundamental value of free enterprise and an endless appetite for products. As David Buxton argues: "The very free enterprise system on which American society is founded has produced a twin, foreign menace; a junk capitalism from the Third-World (especially Columbia) which dumps worthless, destructive commodities on the American market, feeding a never-ceasing demand for the Vice which is destroying society from within" (1990: 149). In *Miami Vice*, the beautiful pastel landscape is built on a swampland, the American Dream is a cancerous drug, and nobody can "just say no."

Other series tended to ignore the issue of urban decay all together. *T.J. Hooker* and *Hunter* are mainly set in non-descript versions of Los Angeles, and *In the Heat of the Night* is set in a fictional small town named Sparta. Within these settings, these series typically forego stories about street-level crime in favor of

more sensational material such as serial killers, mobsters, bad cops, revenge plots, and high-society murders. The final confrontations between the cops and the criminals often occur in spaces outside the city or town, such as an abandoned military fort (*T.J. Hooker*, "The Decoy"), an electrical plant (*Hunter*, "Rape and Revenge"), or a roadhouse (*In the Heat of the Night*, "Intruders").

When concerns about public safety are raised, community-based solutions are often pushed to the side as unnecessary. When the Chief of Police in Sparta is questioned by the town council about how he intends to protect fellow citizens in light of a string of violent assaults on older women, he assures them that he has ordered extra patrols and is personally overseeing a detailed investigation (*In the Heat of the Night*, "Intruders"). When this answer fails to satisfy the most critical council member, a retired police officer from Philadelphia steps forward and begins to extol the virtues of Neighborhood Watch – a program designed to address the fears of inner-city residents. This solution is met with groans from the audience at the meeting, as it clearly does not speak to the needs of this community. The Chief uses the interruption to make a quick exit and get down to the real business of finding the person responsible for the attacks.

The criminal characters in these series are typically psychotic loners prone to violence, greedy opportunists, or simply good individuals who crack under pressure. Bringing these people to justice occupies a central place in each episode, pushing aside any social or political solutions to crime. For example, in the *T.J. Hooker* episode, "Vengeance Is Mine," a serial rapist named Larry Foster is preying on young women. One of these women is the daughter of Det. Paul McGuire, Hooker's former partner and close friend. The daughter identifies Foster by his voice during a line-up, but she is told that the voice identification will not hold up in court. Det. McGuire becomes desperate to bring Foster to justice, eventually taking him hostage. Hooker is forced to step in just before his friend crosses the line by murdering Foster. This focus on individual guilt (in this case, both Foster and McGuire) pushes other kinds of social, political, or economic discourses to the margins of the narrative. Foster's actions are based purely on sadistic pleasure. And while McGuire's breakdown seems connected to his sense that due process rules have prevented justice, Hooker's perseverance in gathering evidence for the District Attorney's office ensures that the case against Foster will be made, thus rendering McGuire's actions premature at best.

As this last example suggests, the effectiveness of the justice system is another issue that is often centrally important to the police dramas of the 1980s. Especially important to these series are the effect of due-process regulations (often understood as "rights of the accused") and the rise of the victims rights' movement. Perhaps the most significant procedural development in law and order in the second half of the twentieth century was the Supreme Court's ruling in *Miranda v. Arizona* in 1966. *Miranda* codified the responsibility of police to inform potential suspects that they had recourse to legal representation before being subjected to interrogation. Should they waive that right, they also had the right to refuse to speak with the police (Friedman 1993: 301). The *Miranda* decision was a source of controversy immediately. Many, including 1968 presidential candidate Richard Nixon, viewed *Miranda* as the logical extreme of the "excesses of the Warren court for 'coddling criminals' and 'handcuffing the police'" by opening the door to wide-ranging complaints against the police for violating the rights of the accused (Malone 1986: 367).

These suspicions swirl outside of criminology and policy circles as well, and are most frequently articulated in popular culture. As early as 1971, Don Seigel's *Dirty Harry* used the restrictions of *Miranda* as its straw man, providing Det. Harry Callahan (Clint Eastwood) with the necessary impetus to move outside of the official legal system in order to apprehend an insane serial killer. Another, more extreme, example of the cultural critique of *Miranda* can be found in the *Death Wish* films starring Charles Bronson. Bronson's character, an architect named Paul Kersey, avenges the brutal murder of his family by working outside of a system depicted as being incapable of meting out proper justice.[13] His gruesome methods of attaining the justice betray the extent to which *Miranda* was seen as unfairly protecting violent criminals. These films seem to suggest that in a society that insists on coddling its criminals and letting them "walk" on technicalities, the only solution is an extremely personalized form of the death penalty. The popularity of *Dirty Harry* (and its various sequels), as well as the *Death Wish* series, indicates the degree to which these views had come to resonate in the larger culture – a change spurred on by news organizations more attracted to stories about crazed and violent criminals allowed to walk among us than to stories about maintaining a justice system that treats even its most violent offenders as "innocent until proven guilty in a court of law."

Stephen Cannell's popular television series, *Hunter*, brought the popularity of *Dirty Harry* into America's living rooms. Sgt. Hunter (Fred Dryer) probably shot more suspects in the line of duty than any other prime-time cop in history. The typical episode of *Hunter* (though not all episodes) culminated in the violent death of the primary suspect in the investigation. In fact, as the series title indicates, the narrative thrust of most episodes was not the laborious process of discovering who bore responsibility for a particular crime (and certainly not the intellectual exercise of determining ultimate ethical culpability for crime), but rather the narrowly focused and literal hunt for the individual guilty of a particular act of deviance. Of course, each episode in the series worked very carefully to establish the suspect as guilty beyond any reasonable doubt. Violent action against this person was made to seem like the logical (and only remaining) result of the narrative. A clear example of this tendency can be found in the two-part episode, "Rape and Revenge." In this episode, Hunter's partner McCall (Stepfanie Kramer) is raped by Mariano (Hunter is later shot), who receives political immunity from prosecution because he is the son of a foreign ambassador. "I think he's gonna get away with it," says McCall while visiting Hunter in the hospital. "Not if I have anything to say about it," says Hunter.

The episode then follows Hunter's quest for revenge against the perpetrator, who has since left the United States and returned to "Curaguay," his home country. McCall, worried that Hunter is ready to step outside the boundaries of the justice system, confronts her partner:

McCALL: Hunter, will you just listen to me for a second? You are a *cop*. You cannot act outside of the law here.

HUNTER: Hey, wait a second. Just hold it just a minute. Outside of the law? Are you kidding me? You know, back in the United States, this guy's committed two rapes, one murder, and one attempted murder on me. Now down here, he's already raped one woman, and his wife's suicide was very questionable. This guy's the one outside of the law, not me.

McCALL: You think I haven't thought about blowing the guy away myself? Do you know how many times I've wanted....Look, nobody wants revenge more than I do, you know, but we took an oath when they gave us these badges. Now I just want this to stop now. I just want this to stop right now. Just let it go.

HUNTER: You think this guy's committed his last rape or murder?

McCALL: All right, well maybe we can think of some other thing here. Um...maybe we can approach his father. He's the head of Internal Security down here. We can confront him...threaten him with going to the press or something.

HUNTER: Wait a minute. The press? Are you kidding me? This is Curaguay. You think there's a press down here?

McCALL: Well, maybe his father doesn't know what a pig Mariano is. Maybe he'll listen to one of his son's victims. You can't just assassinate him.

HUNTER: OK. Go to the General. Go down there and see what he says.

In this exchange both sides of the argument get aired. Even though Hunter appears recalcitrant he eventually succumbs to McCall's argument, though with an air of superiority, knowing full well that the General will refuse to heed McCall's legal version of justice, opening the door to the kind of confrontation he has sought all along.

But when Hunter finally does corner his assailant, he thinks twice about killing him and, as if finally agreeing with McCall, decides to let him go. But here the narrative lets us have it both ways. Just as Hunter turns his back, the suspect gleefully pulls a gun in order to kill Hunter. Hunter anticipates this action just in time, spins around and fires his revolver, killing the suspect. The logic of this scene points in two directions: the system is left intact but is shown to be blind to the real nature of violent criminals. The fantasy of violent justice is enacted under the guise of self-defense – seemingly the last refuge for justice within a system that is seen to coddle criminals.

The perceived coddling encouraged by the *Miranda* decision is closely connected to the victims' rights movement, which began in earnest in the late 1960s as both the civil rights movement and the women's rights movement gained momentum. As Smith, Sloan, and Ward point out, the initial goals of the victims' rights movement were extensions of these earlier movements. "The liberal ideologies of those movements were transformed into a victims' agenda supported by women, minorities, and younger persons, and whose leadership was characterized as drawn from the highly educated, more liberal middle- and upper-income segments of society" (1990: 489). Early advocates for victims' rights often referred to the "second wound": the callous treatment that victims (especially women and minorities) often received at the hands of police officers and defense lawyers. The original goal of the movement, then, was to *protect* specific classes of victims from unequal treatment by the legal system.

This more liberal agenda of the victims' rights movement, however, was slowly challenged throughout the 1980s as the movement grew. Victims' Rights in the 1980s shifted from a *protection* model to a *retribution* model. Diane Kiesel points out that at least some of the increased activity around victims' rights issues might be attributable to "a conservative mood among the populace" and a backlash against "the Warren court's liberal concern about the constitutional rights of the accused" (1984: 25). For example, in order to deliver on his "tough on crime" campaign promises, President Reagan created the President's Task Force on Victims of Crime in 1982. This legislation typically called for tougher standards in sentencing and shifting the focus of victims' rights toward retribution and away from equal treatment within the system (Smith, Sloan, & Ward 1990: 490).

The scene from *Hunter* discussed above is an extreme example of the *retribution* model of victims' rights, but its familiarity – its relationship to so many popular narratives of the time – underscores the potency of the discourse. At the same time, however, the *protection* model remains relevant in the police genre as well. In an episode of *Cagney and Lacey* ("Biological Clock"), Chris and Mary Beth attempt to persuade a reluctant older woman, Mrs. Givens, to testify against her landlord, Mr. Nolan, for his role in the death of one of her neighbors, Mr. Poulianakas. Nolan has succeeded in getting several continuances from the court and it appears that, without the help of this witness, he will get away with the murder. Mrs. Givens, however, is too scared to venture outside of her apartment; she is convinced that she, too, will become a victim.

Within the context of this plotline, she already is a victim. While she has not been assaulted or burglarized, she feels isolated and afraid of a world that doesn't value her. Like the minority and female victims for whom the earlier victims' rights movement was designed, Mrs. Givens does not believe that she has a voice in the system. It is up to the police detectives to convince her that her voice does matter and will be heard. In the penultimate major scene of the episode, Mrs. Givens finally comes forward to tell the story about her neighbor's death and all parties seem satisfied that they, indeed, have a case. Before she can leave, however, the prosecuting attorney asks her why she waited so long to come forward. To this question, she responds: "I'm not a brave woman, Mr. Burke. For the last ten years I haven't asked much of life – just not to be hurt. I was afraid Mr. Nolan would retaliate against me. That's all I could think of. I feel very sorry and ashamed. That's not a way to live." In the end, the episode highlights the importance of protecting victims of crime, even if they are merely *potential* victims.

This episode also emphasizes the important role that citizens play in police work. As criminologist Albert Reiss, Jr. has demonstrated, there is a necessary interdependence between communities and the police who serve them: citizens of any given community act as one part of "the operating system of criminal justice" (1977: 115). The primary role of citizens is to mobilize the police through a call for help or action. This role is often extended through various forms of "community policing." Like the victims' rights movement, community policing encompasses several competing discourses around a core concept. From a citizen's perspective, community policing involves looking out for one's own community through various programs such as Neighborhood Watch and Crime Stoppers, whose primary function is to "observe and report" to the police (Garofalo & McLeod 1989: 328).[14] From the police perspective, community policing involves a greater emphasis on community-based patrol officers who establish a presence and become a recognizable entity within the community itself. These two forms of community policing are often at odds with one another. The police count on citizens to help them keep order in neighborhoods and to help them apprehend suspects. But the police are wary about giving citizens too much power to act on their own in criminal matters. The result is an ongoing tension between the populace and the police around the question of what can be done to protect neighborhoods.

As a set of contemporary discourses, community policing (and the tensions that went along with it) was a fairly regular staple of police dramas in the 1980s. In an episode of *Hill Street Blues* ("Up In Arms"), Capt. Furillo is forced to deal with a group of local merchants who have become frustrated by the lack of police presence in their neighborhood. The "Dekker Avenue Merchants Association" makes its presence felt when a large group of its members arrives in the station house with a robbery suspect in custody. Furillo confronts the leader of the group, Mr. Viatoro:

FURILLO: Since you seem to be in charge here, would you like to give me one good reason why I shouldn't arrest each and every one of you? Talk to me.

VIATORO: OK, this is nothing personal here, OK? I mean, we know how hard it is. But the fact of the matter is, you guys can't protect us. We can't go on the way we are. So we formed an association. This is only a part of us. And we're not leaving this building until that kid gets booked and we get sanctioned by the police department.

FURILLO: OK. That's clear. I hear you.

The initial decision from headquarters is to refuse the group's demand to be recognized as a policing organization. Instead, the group is given information on how to establish a Neighborhood Watch. Mr. Viatoro tells Furillo that the program is "an insult to our intelligence." The association is seeking gun permits and the power of arrest. The merchants have the attention and support of a television reporter, Cynthia Chase, who suggested such action in the first place. She also tells Furillo that the Neighborhood Watch program is ineffective, and that it only appeases naïve citizens. With a television reporter supporting the merchants, the potential for a public relations problem convinces the Chief of Police that the association should, against Furillo's advice, be sanctioned.

This situation in *Hill Street Blues* relates directly to the earlier example from *Hunter*, in which Hunter and McCall argue over the limits of police action. In both cases, the scenes focus on characters caught between their roles as victims and protectors. Both scenes are concerned with the line between the law and vigilantism and both decide that the latter ultimately has its place. The key difference, however, is that *Hunter* constructs it as the inevitable consequence of dealing with an evil individual who refuses to recognize the law. When Hunter kills Mariano, he does so only after refusing pure vengeance. As a result, *Hunter* offers an uncomplicated solution that has it both ways: Hunter abides by his duties as a police officer, but gets his vengeance through the unrepentant action of Mariano. On the other hand, *Hill Street Blues* constructs a complicated dialogue about the virtues and limits of community policing. When Furillo and Cynthia Chase discuss the situation of the Dekker Avenue Merchants Association, they both make compelling points: Furillo insists on avoiding the hazards of arming citizens and giving them the power of arrest, and Chase makes an equally compelling point that the citizens have a right to provide a safe neighborhood for themselves, even if it means acting as the Law.

Taken together, the legacy of urban decay, *Miranda*, the rise of the victims' rights movement, and the increasing attention paid to community policing initiatives in the 1980s laid the foundation for a rich dialogue about the causes of crime and how it should be addressed. Though these issues do not encompass the entire range of narrative possibilities or storylines, they do suggest the degree to which innovations in the police dramas of the 1980s were tied not only to generic developments but also to changing conceptions of the nature of policing itself. The characters and situations depicted in the police dramas of the 1980s embodied the struggle between social discourses both within and between individual series. These social discourses provide the backdrop against which the formal and structural elements of these series operate. In what follows, I address the way that visual style, narrative structure, and character development all contributed to the way these series addressed the social discourses that I have discussed so far.

Style, Structure, Casts, and Characters

The social discourses that I have explored in the previous section are the thematic foundation for the police drama in the 1980s: they are the grist for the mill. In what follows, I want to explore some of the ways in which the mill operates. The 1980s was a crucial period for expanding the language of the police drama. The period saw, or in some cases built upon, several key innovations in the visual style, narrative structure, and characterizations of the television police drama. At the same time, not all producers pushed the envelope in these terms; more traditional representations of the police continued to exist. Additionally, productions outside of the category of fictional drama also influenced the genre in significant ways. Given this range, I will demonstrate that the police dramas of the 1980s not only brought different perspectives to bear on the important issues of the day, but presented those perspectives with an equally diverse array of representational strategies.

The most apparent way in which the language of the police drama was expanded in the 1980s involved matters of style. In particular, the 1980s saw two seemingly opposite (but closely related) developments in visual style: both *toward* and *away from* greater realism. Both of these trends can be identified with what John Caldwell has called "excessive style," or "stylistic exhibitionism" – a self-conscious *performance* of style brought on by a series of changes in the structure and function of the television industry (1995: 5).[15] As Caldwell has argued, in the television industry of the 1980s, "a structural inversion had taken place between the presentational functions of narrative and style" (1995: 67). Rather than the narrative structure dictating the look of a series, by the 1980s "the look of the show frequently organized the narrative" (1995: 67).

The first of these formal innovations – the movement toward greater realism – is represented most clearly by two shows at either end of the decade: *Hill Street Blues* and *COPS*. Both series are marked (at least in part) by "messy" handheld camerawork and a busy, crowded *mise-en-scène* that work together to create an environment of chaos and danger. These aesthetic choices, however, result in drastically different tones, demonstrating that style alone does not guarantee meaning – that form and ideology are not as easily linked as one might suppose.

"Make it Worse": The Realism of Hill Street Blues

Todd Gitlin states that "*Hill Street's* achievement was, first of all, a matter of style" (2000: 274). The show's stylistic achievements were manifested at several levels: narrative structure, character development, casting decisions, dialog, situation, and of course visual style. In each case, the goal was toward greater realism: greater fidelity to the look and feel of "real" police work in the decaying urban environments of the late 1970s and early 1980s. Each episode of *Hill Street Blues* began with the low and busy chatter of a crowded roll-call room over a black screen with only the date and time of day stamped in the lower third of the screen.

The black screen was then immediately replaced by a series of handheld shots of the various occupants of the room as they listened (or not) to the primary voice on the soundtrack: Sgt. Phil Esterhaus, detailing the multiple crimes and concerns that would provide the primary focus of the episode. The look of these sequences was designed to approximate the gritty, random feel of the direct cinema work of Alan and Susan Raymond (*The Police Tapes*) and Frederick Wiseman (especially his 1969 film, *Law and Order*).[16] Additionally, the visual style of the series was intended to compliment the gritty quality of the dialog, the randomness of the situations, and also to capture the texture of feature films that were experiencing some degree of success at the boxoffice (such as *Serpico*, *The French Connection*, *Dog Day Afternoon*, and *Fort Apache, The Bronx*). Gitlin documents this impulse in his conversations with Gregory Hoblit, the first line producer on the show in charge of realizing the look and feel of the series. "I read the script," Hoblit remembers, "and immediately a whole visual sense came to me about what it ought to be. Hand-held camera. Let's get the film as dirty as we can. What I said is 'Let's go for the *Serpico* look'" (Gitlin 2000: 290).

But to speak of realism is to open up questions that go beyond the vagaries and minutiae of production decisions and tales of maverick producers and directors. Bill Nichols argues that "style is not simply a systematic utilization of techniques devoid of meaning but itself the bearer of meaning" (1991: 79). Questions of realism are ultimately questions of representation and ideology. The images we present to and of ourselves say something important about our values and beliefs. Images present worldviews; they are markers of ideology which, according to Nichols, is "the image a society gives of itself in order to perpetuate itself. … Ideology uses the fabrication of images and the processes of representation to persuade us that how things are is how they ought to be and that the place provided for us is the place we ought to have" (1991: 1).

For Nichols, there is a distinct connection between form and ideology. But a close look at another stylistically influential series, *COPS*, will reveal that a similarly messy, chaotic style can speak in a voice quite distinct from that of a show like *Hill Street Blues*. Where the messiness of *Hill Street Blues* underscored the crisis of authority in the police, *COPS* uses similar techniques to *re-establish* that authority. *COPS* provides the viewer with the sense of a voyeuristic experience, but one that is, in the end, carefully guided and narrated by the police who are seen as authority figures in complete control of the situation at hand.

The Voice of COPS

COPS was first introduced to the viewing public in 1989 on the still fledgling Fox Broadcast Network. The series was not the first attempt at marrying reality television and crime: NBC's *Unsolved Mysteries* and FOX's own *America's Most Wanted* represented earlier attempts at the form. But both of these programs looked more like standard dramatic fare, relying on reenactments of crimes already committed and adhering to a standard mode of production that favored continuity in time and space. What *COPS* brought to the table, of course, was "real people and real crime" without actors. Its aesthetic is more in line with direct cinema than the classic realist structures of reenactments. On *COPS*, camera operators ride along with the police on their nightly rounds and capture arrests as they unfold. The camera is left running during chases, creating a vertiginous experience for the voyeur/viewer in the process. And unlike the more "populist" approaches to crime that *America's Most Wanted* advocates, *COPS* presents what Jessica Fishman calls a "progressive myth" in which "official agents are the heroes; their agenda is a rarefied one of advanced technology and specialized knowledge; and the narrative highlights their seemingly immediate successes" (1999: 270).

The raw realism of *COPS* hides a larger strategy of carefully structured continuity. This continuity is provided by a specific editing style that employs the voices of the police officers on the soundtrack to hide gaps in time and space that are the inevitable product of editing hours of footage down into a twenty-three minute television episode. It's no secret of course, that a show like *COPS* is highly edited. A former producer on ABC's *American Detective* has pointed out that reality-based television is shot and edited so that viewers see and hear everything from the police point of view. This unacknowledged but powerful ideological orientation also determines which incidents producers select for inclusion from the hundreds of hours of available footage. Producers simply choose not to include incidents "that make it difficult to determine who the real criminal is and to what extent he or she may be the victim of a system that has careened out of control" (Seagal 1993: 6). In a typical episode of *COPS* the root causes of crime, especially poverty, go largely unrecognized. Instead, viewers see only the situation at hand. The police officer on the scene serves as the sole source of knowledge about the crime, offering his or her understanding of the immediate incident as a sufficient explanation. As Robin Andersen suggests in her book, *Consumer Culture and TV Programming*: "'Video realism' convinces us that we've 'seen with our own eyes,' yet we depend on the cop's sources, perspective and judgment to make sense of the world he defines as criminal" (1995: 176). In *COPS*, this narrative authority granted to the police officer is underscored and supported by the use of their authoritative voices, carefully edited to create continuity within a scene that bears no necessary narrative coherence.

A segment from an episode shot in Palm Beach County, Florida, provides a particularly good example of the cop's function as narrator and the way the authority of his voice is maintained through techniques of continuity editing. As the segment begins, we are located in the backseat of a squad car (as it is with every segment) listening to the driver talk about being a cop. With the camera trained on the driver, the voice of the officer in the passenger side (Corporal Kevin O'Brien) interjects into the conversation. The camera then *cuts* to O'Brien talking. Without two cameras present, the conversation had to have been broken up and O'Brien's voice used as a bridge between the two shots (the driver and himself). This use of his voice gives O'Brien control of the narration and is then used throughout the segment to keep him in control. All information is now filtered through him.

The suspect, once located, runs from the officers who quickly chase after him. The camera operator runs after the suspect as well and the wildly swaying single shot from the camera is left intact, a visual marker that confirms the immediacy of the footage. Once the suspect has been apprehended, the officers begin a search for the drugs that have been scattered in the attempt to escape arrest. O'Brien's voice is now the primary audio element in the sequence as he explains the difficult search for the drugs. His voice once again serves as the unifying element that ties together separate shots. In order to avoid jarring jump-cuts, the editors typically splice the audio segments when the camera has turned away from O'Brien in order to hide the audio edit. The result of this technique is a unified narration that implies his mastery over the situation.

The editing strategies of *COPS* maintain the authority of the narrating officer through the creation of a unified space out of fragmented shots. The constant movement of the camera, continually reframing through pans, tilts, and zooms, works to disorient the viewer by referring back to the uncut handheld chase near the beginning. The fact that the officer's statements seem continuous but are, in fact, cut up by different shots of both the surrounding area as well as shots of the officer in different physical locations within that area, points to the degree to which this program adapts the visual cues of *cinéma vérité* in order to create a narrative of containment.

There is a clear narrational presence in *COPS* in the person of the primary officer on the scene. Furthermore, while there are no overt editorial devices, such as dissolves, the use of audio edits serve the same purpose. Most viewers would notice a dissolve as something other than the straight recording of continuous time and space, but audio splices are difficult to notice without close and repeated viewing. They are able to establish relationships between shots, which can then pass more easily for transparent reality. More importantly, this way of constructing the soundtrack necessarily privileges the police officers' voice as the voice of authority. The officers in *COPS* become the active *subjects* of the narrative, through whom all pertinent information is filtered and whose mastery of time and space allow them to define the situation in terms that are most decidedly their own.

Away from Reality

At the same time that *Hill Street Blues* and *COPS* were moving *toward* an enhanced realist aesthetic (though toward different ideological ends), shows like *Miami Vice* moved in the opposite direction: toward a more expressionistic style. As discussed in the previous chapter, the aggressive style of *Miami Vice* was first and foremost a product of the economics of the television industry in the 1980s – a matter of distinguishing the series in a crowded marketplace, and capitalizing on current cultural trends. But beyond economics, the style bears a relationship to the stories being told. But just as the examples of *Hill Street Blues* and *COPS* demonstrate that there is no necessary or singular correlation between form and ideology at the level of production, *Miami Vice* illustrates the fact that the relationship between form and ideology varies along critical lines as well.

The style of the series is expressive of the kind of circles that the cops travel in. As undercover vice detectives, the characters have to be able to pass as high-level drug dealers, with all the accouterments that the lifestyle entails. But at the same time, the excessiveness of the series' production design seems to betray a certain commitment to consumer culture: this is commercial television, after all. Some critics saw the series' lavish and surprising style as pushing the narrative toward the realm of more socially critical texts, perhaps in the way that the films of Douglas Sirk are often considered to be critical *in* their excess. Other critics see the style as a potentially dishonest distraction from the real ideological force of the series.

Jeremy Butler has argued that *Miami Vice* brought the style and sensibilities of film noir to broadcast television. In particular, Butler argues that "*Miami Vice* shares at least three principal themes with *film noir*: moral ambiguity, confusion of identities, and fatalism (caused by a past that predetermines the present)" (1985: 130). Because the action in *Miami Vice* involves undercover work, the moral ambiguity and confusion of identities are bound up with police work itself:

> The form of detection that goes on in *Miami Vice* owes less to Sherlock Holmes-style ratiocination than it does to the *film noir* and, more generally, American hard-boiled fiction, in which the private eye is implicated in the crime he is supposed to solve. The policework in *Miami Vice* is based on masquerade – bordering on entrapment – rather than well-reasoned deduction.
>
> *(Butler 1985: 131)*

Because the rationalism of the classical detective story and the police procedural are largely missing from the highly stylized world of *Miami Vice*, Butler argues that the show works "against the classical narrative model" and offers the viewer "a pleasure that is not normally available on television: the pleasure of gazing, of considering the image as image" (1985: 137). For Butler, this stylistic strategy of *Miami Vice* contains within it the potential to disrupt easy identification with the image, and confront the viewer with a more complex and morally ambiguous vision of law and order, especially as it is traditionally represented on television.

David Buxton, on the other hand, has argued that the highly stylized *mise-en-scène* of *Miami Vice* actually leads *away from* any confrontation with moral ambiguities. For Buxton, *Miami Vice* represents two key pillars of Reagan's America:

a vision of justice that is located squarely on the shoulders of morally corrupt individuals, and the celebration of conspicuous consumption:

> The series seeks a terrain on which they can co-exist without coming into contradiction, in order to circumvent the claims by political critics of economic liberalism that it is precisely the perverse social effects of the latter (consumption) which exacerbate an already serious crime problem in late capitalist societies. In the series generally Reaganian options, endemic crime and economic crisis are the fault of moral weakness (Vice); on the other hand, there is nothing (morally) wrong with conspicuous consumption and the pursuit of private wealth.
>
> *(1990: 142)*

For Buxton, it is the obsession with style and display that work to distract the viewer from the contradictions of the ideology at work and to offer, instead, a world undeniably pleasurable in its aesthetics of abundance.

Whether the style opens up the possibility for dealing "realistically" with the complexity of crime and punishment (as Butler might argue) or distracts from the "real" causes and consequences of crime in Reagan's America (as Buxton would argue), there is certainly evidence on both sides. What emerges from their debate is the point that style cannot be read in any singular way: style and technique are a set of tools with multiple uses and functions. The relationship between form and ideology needs to be considered in the context of individual series, and as the preceding examples illustrate, even that context may be read differently by different critics.

Serial Narratives

Of the range of innovations in the police genre with which *Hill Street Blues* is often credited, the most notable is the turn toward seriality. This innovation within the genre, however, took place within a larger move toward serial drama across the television networks. In fact, looking at the 1980s as a whole, Jane Feuer has argued that "serial form was the aesthetically dominant narrative innovation of the decade" (1995: 113). The daytime serial – or soap opera – has been a staple of network programming since the early days of commercial radio. The prime-time serial has roots as far back as the 1960s serial, *Peyton Place*, and the first prime-time serial to achieve status as a bona fide hit was *Dallas* which first aired in 1978. One question that we need to ask about seriality in prime-time, then, is whether or not it functions in the same way for all series based on a continuing narrative. In other words, does the formal and structural innovation of seriality in the police genre serve specific functions that can be recognized across series? If not, what then are the differences and why are they important?

Prime-time soap operas were still a relatively novel form when *Dallas* first emerged, and the structural emphasis on seriality took a range of different forms as serial dramas began to proliferate. While *Hill Street Blues* has often been understood as a hybrid of the police genre and the soap opera, there are clear and important distinctions to be made between the more soap-operatic serials such as *Dallas* and *Dynasty* (and their various progeny) and *Hill Street Blues*. *Dallas* and *Dynasty*, of course, followed the structure of the daytime soap opera fairly closely: the narrative followed a string of constantly evolving interpersonal relationships and professional machinations that never seemed to resolve completely (and certainly never quickly). *Hill Street Blues*, on the other hand, is an example of what Thomas Schatz has called a "semi-serial format" in which some primary story lines are resolved within a single episode, or often over the course of two or three episodes, while secondary storylines (often personal in nature) are allowed to develop over the course of several months, seasons, or even the entire run of the series (1987: 90).

Schatz points out that *Hill Street Blues* was not so much a clean break with television's past as it was a further development of an existing trend in programming. Far more influential for Schatz was MTM's first drama, *Lou Grant* – a spin-off from the situation comedy, *The Mary Tyler Moore Show*. *Lou Grant* built upon the already established trend toward an ensemble cast that "coalesced into an integrated constellation of characters from disparate backgrounds, working together, whose commitment to their work and to one another had become the governing force in their lives" (Schatz 1987: 88). This ensemble format was developed most successfully by the situation comedies, *The Mary Tyler Moore Show*, *Barney Miller*, and *M*A*S*H*; but these series had not developed the semi-serial format that would come to define *Lou Grant*. For Schatz, "*Lou Grant* was less a clone of its predecessor than an experiment in generic recombination, with its unique blending of comedy and drama, of realism and stylization, of episodic and serial story lines, of social relevance and soap-opera melodramatics" (1987: 86). Most of the MTM hits of the 1980s, including *Hill Street Blues*, followed in *Lou Grant*'s footsteps.

For the police drama, however, *Hill Street Blues* did represent a significant development within the form. While the police drama has always been predicated on social relevance for its narratives (storylines pulled from real case files or "ripped from the headlines"), and there had already been a clear trend toward melodramatic characterizations of the police that focused on their personal lives, *Hill Street Blues* significantly expanded these trends through the elements it borrowed from *Lou Grant*: an ensemble cast and a semi-serial narrative structure. The cast of *Hill Street Blues* included as many as thirteen regular recurring characters and during its first season a typical episode balanced at least four different story arcs, "each starting at a different moment and often not resolving at all" (Gitlin 2000: 274).

Though *Peyton Place* in the 1960s and *Dallas* in the 1970s had already established the potential for a prime-time serial to have significant popular appeal, NBC fretted over *Hill Street Blues* and its open-ended narratives. As Tom Stemple illustrates, one of the stipulations made by NBC when it renewed *Hill Street Blues* for a second season (after nearly disastrously low ratings in its first season) was that "there be at least one storyline completed per episode, since the network felt audiences had trouble following storylines wandering through every episode" (1992: 229). Given the existing precedents (and the runaway popularity of *Dallas*), the network's demands about closing off at least one storyline on *Hill Street Blues* perhaps suggest more about genre and ideology than audience dissatisfaction: stories about crime and those who work to solve those crimes demanded some sort of resolution.

For many critics (Deming 1985; Zynda 1986; Thompson 1996), the serial structure of *Hill Street Blues* represented progressive potential for the series. Rather than resolve individual cases in the space of an hour, serial narratives represent the opportunity to explore the complexities of crime and investigation, engage with social contradictions, and chart change over time. But Jane Feuer warns that the notion of "progress" that serial narratives represent (i.e. a storyline that continually moves forward, with characters that change over time) should not be confused with the political sense of the word. In fact, for Feuer, characters in serial narratives do *not* necessarily change. Rather, they "perpetuate the narrative by continuing to make the same mistakes" (1995: 128). Instead, within a multiple plot structure, difference and change are manifested in the more simple fact that "characters' positions shift in relation to other characters" (1995: 128).

Seriality then, at least in part, dictated that the story emphasis be placed more firmly on the personal relationships of the police, rather than their professional activities. As Steve Jenkins points out, the melodramatic elements of *Hill Street Blues* typically outweighed the more procedural elements of the traditional police drama by focusing on interpersonal relationships through "carefully structured arrangements of characters, confrontations, encounters, incidents, and situations, which constitute the body of each episode" (1984: 194). The real thrust of the world of *Hill Street Blues*, then, is not so much the individual crimes that are being dealt with, but the way in which those crimes activate a set of relationships that are illustrative of larger social relationships.

Characters and Casts

Since seriality was particularly important to character development over time, we need to examine the types of characters that emerged during this period. When we consider the kinds of choices that producers and programmers make about types of police characters, one of the central issues is how the character is positioned in relation to heroic action. Typically, the character of the police officer

has been predicated on a range of heroic abilities, including ph
finely-honed instinct, wisdom gained through experience, a healt
police bureaucracy (bordering on disdain), a sense of selflessness,
to one's peers (especially one's partner), and an intimate knowledge of criminal circles. Most police dramas of the 1980s included central characters with many of these traits, but they also expanded heroic action to include emotional bonding and social sensitivity as well.

The degree to which emotion and sensitivity are seen as heroic, however, varies across series. *Hill Street Blues* typically explored the impact that the inherent danger of police work (physical and emotional) had on the officers' relationships with one another. For example, the first season of *Hill Street Blues* explored the struggles of Bobby Hill and Andy Renko to maintain their close relationship after nearly being killed in a random shooting (an incident that took place during the episode discussed earlier in this chapter). The investigation of the shooting does continue over the course of the first several episodes, but the real impact of the shooting is felt within the increasingly tense relationship between Hill and Renko following their release from the hospital. This tension culminates in a scene from the series' third episode ("Politics As Usual") in which Capt. Furillo grants the partners a "divorce" until they prove that they can get along on the job. The realization that they will be split up brings the two men to a tearful reconciliation:

RENKO: You wouldn't last more than ten minutes out there without me, anyway.
HILL: Oh yeah? I haven't been doing too good with you, anyway. *(The two begin to laugh instead of cry. Renko stands up and embraces Hill.)*
RENKO: We need each other, Bobby. *(Renko wipes Hill's tears)*
HILL: I'm afraid we do, cowboy.
RENKO: Look, why don't we just go get something to eat and go back to work.
HILL: Oh, God.

Clearly, while Furillo's goal is to get the two men back on the job, the scene itself is constructed to underscore the important emotional bond between the two men. The ability to communicate and express their emotions is the only thing that can save their partnership, and maybe even their jobs.

More "traditional" series like *T.J. Hooker* and *Hunter* managed this issue in very different ways, emphasizing violent (masculine) solutions to crimes. In the episode of *Hunter* ("Rape and Revenge") discussed earlier, after Hunter and his female partner, McCall, are attacked by the same man (McCall was raped, Hunter was shot), McCall visits Hunter in the hospital and tells him that the perpetrator is probably going to get away with his crimes because he has diplomatic immunity. The sequence that follows McCall's visit to Hunter in the hospital depicts the two characters "recovering" from their respective injuries. In alternating scenes, McCall is shown talking to a therapist, gradually moving over time from angry tears to a reluctant smile. Hunter, on the other hand, is shown in isolation, working out with weights, regaining the strength in his injured shoulder. The sequence concludes with McCall completely removed and Hunter alone in the desert taking target practice with his pistol; her emotionalism has no place in his version of heroic action. His aim is poor until he notices an eagle flying overhead. Reminded of the importance of his instincts, Hunter resumes his target practice, now hitting every target with ease. The scene suggests rather bluntly that the true responsibility for justice rests with violent, masculine authority.

Between the emotionally capable partners and the violent action hero, a third type of police character emerged in the 1980s: the pragmatic leader who can balance competing ideas about how to handle crisis situations. Capt. Furillo from *Hill Street Blues* is perhaps the primary example of this type of character. Furillo is often caught between three extremes within his own department: the militarism of Lt. Howard Hunter, the sensitive liberalism of Lt. Henry Goldblum, and the wishy-washy political opportunism of Chief Daniels. In the pilot episode of the series, one of the pivotal strands of the narrative involves two young Puerto Rican gang members who take hostages in a liquor store robbery and fall into a long standoff with the Metro police. Det. Henry Goldblum, the precinct's liberal conscience and negotiator, initiates contact with the boys via telephone. Goldblum tries to "connect" with one of the boys and is quickly rebuffed:

GOLDBLUM: Maybe you could tell me what's troubling you?
HECTOR: What's troubling me? What's troubling me is you, chump! (hangs up phone).
GOLDBLUM: (hanging up phone) How am I supposed to create an ambiance of trust under these conditions?

The scene then cuts to Capt. Frank Furillo trying to talk Lt. Hunter out of using his military might to, in Hunter's terms, "neutralize that liquor store…make an example out of it." In the end, Furillo agrees to let Hunter and his men go into the area, but only if they maintain a low profile. What emerges from this brief scene is one of the program's central tensions, which involves the balance between two opposing discourses about handling crisis situations: Goldblum's liberalism versus Hunter's militarism, both of which are played mostly for laughs and rendered largely ineffectual in their attempts to correctly read and react to these situations. During the course of this scene, Hunter's militarism is pathologized and Goldblum's liberalism is feminized (underscored by Hunter's question to Furillo when he fears that Goldblum's perspective might win the day: "It's a gut check, Frank. You wouldn't want a bunch of daisies where your cinch belt ought to be, would you?"). Furillo's role is to manage the situation rather than any particular agenda.

The scene described above illustrates a range of character types as well the hierarchical relationships between those characters and the discourses that they represent. These discursive and dialogic elements of the scene are the product of an ensemble cast. The move toward ensemble casts creates several possibilities. First, on a purely functional level of storytelling, it allows the writers and producers of the series to disperse the narrative emphasis across a range of characters, providing diversity in the storytelling and allowing a range of viewer identifications to develop over time. Second, on an ideological level, this dispersed and diverse cast allows for more complex reactions to (and interactions regarding) specific situations. These reactions and interactions can be explored across political lines of race, class, and gender. Finally, the ensemble cast also has an institutional function. Because the writers and producers can disperse the narrative across multiple and diversified characters, the narrative possibilities of the series necessarily expand and increase the potential length of the series run (provided, of course, that it is popular enough to attract a significantly large and/or valuable audience in the first place). Moreover, multiple characters open up more possibilities for spin-offs and thus increase the chances for developing a popular franchise that can be exploited for even longer periods of time than an individual series.

But what is more important than the simple expansion of the trend toward serial narrative and ensemble casts is the idea that characters could (and would) change over time. Jane Feuer has argued that the combination of serial form and ensemble cast together can foster a more "progressive" narrative environment "in that they affirm bourgeois notions of character development and growth" (1995: 127). This observation is important in that it points to the move away from the traditional police drama's focus on the *professional* dimensions of crime, community, and citizenship. Instead, the serial narrative based on interpersonal relationships and character growth moves toward the *personal* dimensions of these concepts. Certainly, *crime* still exists in a near absolute form (the law does get broken in recognizable ways), but it is the *impact* that these crimes have on individuals and communities that receive the lion's share of the narrative attention. In other words, while the basic elements of crime, community, and citizenship remain at the center of the narratives, the *discursive weight* has been shifted from the professional to the personal (and interpersonal).

In an episode of *Cagney and Lacey*, titled "The City That Bleeds," a young black man has been killed by a gun belonging to a police officer. Though the officer in question did not shoot the young man (his gun was stolen), the incident has divided the city along race lines with regard to the issue of police violence against black men. And the squad room has followed suit. Nearly every scene of the episode provides characters with the opportunity to discuss their perspectives on the issue of race relations. In an early scene, the ethnically diverse squad reacts to a prominent black lawyer speaking about the incident:

LACEY: She was all over the *Post* the morning. She blames the Johnson homicide on the Bernhard Goetz verdict.
CAGNEY: Oh, please!
PETRIE: It's not that far off.
COLEMAN: Well, if the boys in Queens don't solve this soon the mobs will be in the street 'til Chanukah.
CAGNEY: With Miss wanna-be D.A. leading the charge.
ESPOSITO: Wait a second. You should listen to what she's saying. It's been a week and there are no witnesses, no suspects, and Bobby Johnson didn't even have a record.
CAGNEY: Uh huh. A regular Boy Scout.
PETRIE: You've been on the job too long, Cagney. Good kids *do* get wasted, you know.
LACEY: Second black kid killed in that neighborhood this year.
CAGNEY: What was he doing there?
PETRIE: Maybe he thought it was America?

The scene sets up the squad room along racial and ethnic lines and establishes a range of reactions to the incident. These reactions are then explored in more detail (and with growing anger and frustration) as pairs of characters interact. What is important about the construction of this episode is the fact that the crime itself initiates a story not primarily about the apprehension of a suspect, but about the impact of crime on interpersonal relationships and personal belief systems.

While ensemble series such as *Cagney and Lacey* can explore a wide range of relationships among the different police officers, more traditional series such as *T.J. Hooker* and *Hunter* typically work over the same relationships across each episode. In the case of *T.J. Hooker*, this relationship is between the student and the mentor. Each episode of *T.J. Hooker* builds upon this relationship by providing a seemingly endless stream of instructional opportunities: scenarios where Hooker can teach his young charges the proper ways of reacting to, and resolving, danger in their midst. Similarly, each episode of *Hunter* works over the relationship between the male and female partners (Hunter and McCall). As illustrated earlier, the interactions between the two typically take the form of arguments regarding the nature of justice: a distinction between benevolent (legally based) justice and violent (personally based) justice. Importantly, while these training sessions and arguments do divulge personal information about the characters, they are predicated largely on procedural elements about what is and what is not acceptable or effective *police* behavior.

In the 1980s, the range of character types for the police officer was expanded considerably. In addition to the already full stable of tough-minded moralists and anti-establishment loners, the police series of the 1980s often included a wide range of newer characters who brought an expanded sensibility to the idea of what heroic action entails. Women and minorities were regularly included in casts, especially ensemble casts, in keeping with changes in the structure of police departments across the country. But these characters also brought with them a different range of sensibilities and social experiences against which crime could be understood. Furthermore, the emotional lives of characters became increasingly central to the narrative. No longer were emotions reserved for scenes away from the precinct; now interpersonal relationships between cops were part of the drama. Additionally, more liberal minded and socially sensitive cops found their way into squad rooms, though typically counterbalanced by an equally conservative or perhaps just pragmatic foil. This expanded range of character types, placed in various relationships with one another, often in large casts, opened up the possibility for more (and different) kinds of dialog about crime and its impacts than had previously been available.

Conclusion

The decade of the 1980s was a crucial period for adding to the language and conventions of the television police genre. Conventional depictions of police no longer seemed to fit the rapidly changing social and cultural context of the time. A range of social problems and issues put pressure on producers and programmers to develop more relevant and meaningful versions of law enforcement. The rapid decline of inner cities, the civil rights movement, changing notions of the rights of the accused, competing discourses about victims' rights, and problems with police corruption are just some of the issues that dictated the thematic terms for the depiction of police in the 1980s.

Popular culture responded to these issues in many different ways. Authors like Joseph Wamabugh and Ed McBain pushed the police procedural novel in new directions; films like *Dirty Harry* and *Serpico* took positions on issues of police corruption and due process. Television also responded by introducing new styles and characters into the repertoire of the genre. New narrative strategies, such as ensemble casts and serialization, led series to explore interpersonal relationships between cops more readily. These strategies also gave producers a wider range of conflicts and perspectives to explore within the context of a particular issue. Additionally, a wider range of styles was introduced. While police dramas have almost always had some investment in both realism and spectacle, new technologies (such as the video camera) and new social forms (such as *cinéma vérité* and music videos) increased the expressive range of the genre. All of these stylistic developments were put in the service of the kinds of political and social issues that formed the thematic content of the series.

Several key series were influential during the decade. *Hill Street Blues*, *Cagney and Lacey*, and *Miami Vice*, for instance, each propelled the genre in directions that it had only begun to explore in previous years. At the same time, more conventional examples of the genre continued to exist on the airwaves: for example *Hunter*, *T.J. Hooker*, and *In the Heat of the Night*. These more conventional series added to the cultural forum by forming a dialog with more stylistically innovative and socially critical series. Even a stylistically innovative and daring representation of police work such as *COPS* could promote a more conservative ideology, highlighting the fact that there is no necessary correlation between form and ideology.

These issues and ideas are the framework for understanding the genre at the beginning of the 1990s. The dramas that form the case studies that follow in the next three chapters entered into the genre within the context of the changes outlined in the present chapter. They adopted some innovations, rejected others, absorbed innovations from related areas of popular culture, and reinvigorated elements from the genre's past. In each case, these series proceeded with (and should be approached with) a keen awareness of the genre's history.

4

STOP MAKING SENSE

Reflection, Realism, and Community in *Homicide*

In the previous chapter I examined the complex and sometimes contradictory ways that the police dramas of the 1980s expanded the language and conventions of the television police genre. Each of these series responded to changing social, cultural, and political conditions, specifically the rise of the New Right and changing discourses about law and order: concerns about due-process, victims' rights, and community policing. These series were also products of a changing television industry and their responses were registered not only in the thematic concerns across different series and within specific episodes, but in the narrative and stylistic choices that producers made as well: more daring visuals, open-ended plots, and more complex and ambiguous characters. These developments constitute the primary framework for understanding the major series of the 1990s: *Homicide*, *NYPD Blue*, and *Law & Order*. As the next three chapters will illustrate, each of these series represented a critical point in the ongoing development of the police drama. In each case, these series were produced with a keen awareness of the genre's past, through which they opened up and explored specific core elements of the genre: realism, melodrama, and issues of justice. Each series adopted and developed certain thematic and stylistic innovations within the genre while also revitalizing other, more traditional generic elements. It is within these dynamic combinations of innovative and traditional elements that these series participate in the ongoing forum about crime, community, and citizenship that is so central to the police genre.

Perhaps more than any other police drama of the 1990s, *Homicide* took the very idea of telling stories about crime and community as one of its central themes. This chapter argues that *Homicide* represents a critical commentary in the evolution of the police genre in two ways. First, it calls attention to the visual conventions of realism by foregrounding and highlighting unconventional stylistic strategies (such as jump-cuts, overlapping and repeated action, handheld cameras, and canted angles). This approach to televisual style serves a couple of functions: it creates a visual dynamism in a series that was committed to moving away from car chases and gun battles as part of its narrative world; and it simultaneously engages in a highly self-reflexive critique of the generic commitment to realism. The second way that the series engages in critical commentary is by extending the practice, initiated by *Hill Street Blues*, of transposing the narrative of the police drama from weekly resolution of individual crimes to a complex discourse about the social, economic, and political policies that sometimes foster crime in the first place. This narrative approach offers a critique that is less about the genre in particular and more about the social structures, particularly as they apply to the connection between notions of *community* and practices of criminal justice within which police dramas have been produced and made meaningful historically. In these ways, *Homicide* engages in a practice of *commentary* on the genre, questioning what it means to represent crime and police work in late 20th century urban America.

The narrative and stylistic structure of *Homicide* highlights an acute awareness of the tension between reality and its representation via the video image. Whether it was the harrowing found-footage of a series like *Real TV* (1996–2001), which revels in displaying only the most excruciating examples of human folly; the grisly, untamed spectacle of *When Animals Attack* (1996); or the *vérité* stylishness of *COPS*, a vast market had opened up for a generation of video voyeurs whose sole purpose seemed to be the speedy delivery of the chaotic world outside into our living rooms. These televisual thrill rides trade on a fascination with liveness – a fascination that is extended by the proliferation of twenty-four hour news outlets. But they can also serve as sites to interrogate the tension between real events and their media representations. *Homicide*'s use of visual codes that signal reality and immediacy are, in part, a comment on the dubious claims to authenticity offered up by the documentary realist strategies of these other shows: the ways that they function as deliberate constructions of reality, often resorting to traditional narrative strategies and assumptions to tell their stories.

This critical commentary on the codes of realism is also connected to a critical engagement with the notion of community. Through being set in the very real location of Baltimore, Maryland, *Homicide* engaged in a dialog about the way that competing concepts of community are articulated in a specific time and place. Shot entirely on location, the series featured Baltimore in ways that most television series (even *Dragnet* with its opening sequence that highlighted Los Angeles) had not attempted – not merely as a backdrop or a series of identifiable locations, but as a character itself with a personality, a set of specific traits, and a history. At the same time though, Baltimore was able to act as a metaphor for urban America in general. The very particular palette of Baltimore allowed the producers to tell stories that could also resonate with the larger issues facing urban America in the 1990s. Baltimore, like so many cities in the U.S., has undergone rounds of gentrification and urban renewal that have simultaneously revived the economy of the downtown and pushed the city's poorer, mostly black citizens toward the margins. *Homicide* engaged with the effects of this gentrification and a lopsided economy on the various communities of Baltimore in very direct and honest ways, even as it packaged them within the constraints of a one-hour commercial television drama and offered them up for consumption.

What was *Homicide*?

Homicide aired regularly on NBC from January 1993 through May 1999 and was an ensemble drama built around a group of detectives in the Baltimore police department's homicide unit. The series was shot entirely on location in Baltimore. Through a unique blend of documentary realism and the pop art stylishness of MTV, the series challenged many of the central tropes of the cop genre: narratives that tend toward the reaffirmation of clear moral authority; physical action as a solution to crime; and a reliance on a narrow set of realist conventions. But *Homicide* was also, of course, a product of the entertainment industry and, as such, was constrained by certain pressures and practices that would shape its look and narrative structure. *Homicide* emerged at a time when network television was looking for ways to distinguish itself amidst the challenges of cable, and the series producers were initially encouraged to build on some of the innovations of series like *Hill Street Blues* and *Miami Vice*.

The series was adapted from the true-crime book, *Homicide: A Year on the Killing Streets* (1991) by David Simon, a former crime reporter for *The Baltimore Sun*. Simon followed the homicide unit of the Baltimore Police Department for a year and his book is a thick, ethnographic account of the culture of detective work in one of the nation's murder capitals. Simon's goal was to move past the objective style of reporting so central to contemporary journalism and offer an insider's view of detective work:

> To tell the story of a year in the life a big-city homicide unit, I did something journalists seldom do anymore – with crime stories or any other fare. I made a conscious decision to write the narrative from the point of view of the central characters: four detectives, two sergeants, and a lieutenant in charge of the shift. As a result, the reader travels through a homicide detective's daily routine accompanied by a narrator who is, in effect, the communal voice of the homicide unit rather than an overtly detached reporter.
>
> *(Simon 1995: 37)*

This approach revealed not only the difficult – often painful – job that these detectives do each day, it also cast light on more surprising elements of the job: morbid corpse-side humor, casual racism, sexism, and homophobia.

Several reviewers found these inclusions disturbing and claimed that Simon had shirked his duties as a reporter by opting for sensationalism over objectivity. But Simon contends that he was simply after a more truthful illustration of police work – one that pulled no punches and offered readers a glimpse into the lives of the detectives without capitulating to the impulse to offer a moral lesson:

> When I reported and wrote *Homicide* I was fully aware that my detectives were, at moments, racist and sexist and homophobic. And I was also conscious of the fact that their squad-room humor was often little better than cruel banter. I very purposefully kept those moments in the manuscript, left them in every chapter as guideposts for readers seeking an honest view of inner-city cops.
>
> *(1995: 4)*

This tension between honesty, objectivity, and sensationalism forms the foundation of Simon's entire approach to his book; it also provides a rich context for understanding the production of the television series. The series producers picked up on the flavor of Simon's book – both in style and characterization – in order to tell a more "honest" story about police detectives, but were under constant pressure from the network to dull its rougher edges and provide more widely palatable characters, and more action.

Homicide was first introduced to the viewing public immediately following the 1993 Super Bowl – a strategy designed to announce the series as a significant television event. NBC's initial promotion of the series emphasized its cinematic pedigree: namely, the presence of Barry Levinson as one of the executive producers and director of the pilot episode. In addition, screenwriter Paul Attanasio (*Quiz Show, Donnie Brasco*) was commissioned to write the pilot episode and was credited as the series creator. The correlation between Hollywood and *Homicide* continued to be fruitful over the years as an impressive number of feature film directors and actors appeared both behind and in front of the cameras. The series also boasted impressive television credentials, particularly in Tom Fontana. Fontana, already a veteran writer and producer whose previous experience included NBC's critically acclaimed drama *St. Elsewhere*, was brought in to oversee the daily operations of production, and emerged as the major creative force behind the series as a whole.

Importantly, *Homicide* was a show as much about its victims as its heroes. One of the central visual elements of the series was "the board," a dry-erase board on which the names of murder victims were listed underneath the primary detective's name. Names in red ink indicated investigations that were still open; the names were changed to black when the case was closed.[1] As Giardello explained to Bayliss in the first episode, "You look up there, you know exactly where you stand." But as an element of *mise-en-scène*, the board assumed a much greater thematic role than the simple accounting of job success or failure. A central visual trope of the series was an extreme close up of an anonymous hand writing a victim's name in either red or black. This simple image represented an almost compulsive act of naming within the series – a way of "speaking for the dead." Against the harsh white background of the board, the name alone fills the screen. The primary function of these scenes is a brief but telling reversal of the generic emphasis on the central cop/hero. These shots put the victim at the center of the narrative and provide a kind of equilibrium where the impulse to identify with the detective/hero is balanced by the visual demand that we also acknowledge the victim.

Since the series was set in Baltimore – a "brown town" – the names of these victims were often members of the black and poor "underclass" that historically make up a large proportion of the criminals central to the police genre. But, as I have pointed out elsewhere, "rather than revel in their untimely demise, the series worked to place their lives and deaths in the larger context of a society at odds" (Nichols-Pethick 2004: 123). As Christopher Campbell has suggested, *Homicide* was particularly adept at "leaving audiences to grapple with the subtleties and impact of contemporary racism" (2000: 26) – a quality it inherited from David Simon's own misgivings about the relationship between media representation (especially journalistic practices) and justice. As an example of how contemporary racism is perpetuated through the news and entertainment media, David Simon provides a vivid illustration of the journalistic practices that define these victims in particular ways:

> I used to cover crime on the late shift in Baltimore for *The Sun*. It was a living measured, by and large, in four paragraph installments. You'd call the cops, ask what was going on, and then, when they emitted a handful of facts about which body fell on which corner, you'd write it up briefly and send it to the night editor. West Baltimore, East Baltimore, lower Park Heights, Cherry Hill – the rowhouses and postwar housing projects were all decidedly similar, and assuming the casualty was poor and black, the newspaper accounts were similar as well.
>
> *(1997: 18)*

Like the housing projects of Baltimore, these newspaper accounts are nondescript, devoid of any of the outrage and panic that might accompany the story of someone "unlucky enough to be killed in the right zip-code" (Simon 1995: 35). What Simon takes issue with is the "bloodless" reporting that participates in the "devaluation of black and Hispanic life in our cities" and has made inner-city violence "so common and so certain that we can no longer regard it as unusual" (1995: 36).

As an adaptation of Simon's book, *Homicide* was a response to this kind of journalism as well as a response to the similarly uncomplicated equations of race and crime found on television – both in crime dramas and on the evening news – with its penchant for spectacle over reality.[2] But as a commercial product – a police drama made for television – *Homicide* was subject to the pressures of prime-time entertainment: namely, the need to make "gritty realism" palatable for advertisers and a television audience. Simon argues convincingly that, because of these pressures and the standard practices of television production, the series based on his sometimes brutally honest book – and for which he eventually became a regular writer and producer – was guilty of its own kind of misrepresentation of reality. "Cheap motels, broken rowhouses, projects street corners – the same rotted Rustbelt terrain that had yielded all those stunted news briefs suddenly became a studio backlot" and subject to the kind of manipulation and simplification that turns the brutality of inner-city life into a spectacle for consumption (1997: 18). As an example of this simplification, Simon cites the production of a scene from the second season episode, "Bop Gun," in which real drug dealers were used as extras:

> Boo stayed on the corner that day, slinging blue-topped vials of coke. The rest followed Tae across town to the Perkins Homes, a squat stretch of public housing that would serve as the pretend drug market. They filled out tax forms, waited out the inevitable delays and were eventually escorted by an assistant director to a battered side street. There, on the set, a props man handed them pretend drugs and pretend weapons, and the director, a very earnest white man, arranged them on the street in the manner most pleasing to the camera.
>
> *(1997: 19)*

When one of the men in the scene finally complained (after nearly a dozen takes of drug dealers scattering from the street corner as police cars descend upon them) that this wasn't how they would do things in real life, the Assistant Director responded simply: "It's just television, guys" (Simon 1997: 19). The particular scene filmed that day looked, in the end, nothing like reality; but it made for an excellent television moment – a roundup of suspects with a slightly tougher, more believable sense of street life than more traditional cop shows.

This sense of street life was achieved mostly through the visual style of the series, which took a complicated approach to the realist conventions so important to the police drama. According to Tom Fontana, "(Barry Levinson) wanted it done in a documentary style, using editing techniques like jump cutting or repeating the same shot three times, to give the audience the same sense of stimulation as gun battles and car chases, but more in line with the kind of stories we were trying to tell" (Fontana 1998). Fontana highlights the seemingly contradictory fact that the visual approach of the series was designed to approximate both the effect of spectacle *and* documentary realism, all in the service of stories that were different from the more traditional, action-oriented police series. The initial impulse to displace the action onto the camera and the editing was eventually extended into the thematic concerns of the series. As I will discuss in more detail later, this overt style was often used self-reflexively to raise fundamental questions about realism and representation, and to call attention to the highly constructed quality of all media images.

Homicide and the Limits of the Real

An early review of David Simon's book made the point that "if Simon's book were a movie, it would be a documentary" (Uviller 1991: X4). The reviewer was referring primarily to the fact that the book was a work of non-fiction. But he was also referring to Simon's direct, unflinching style in presenting these detectives to the reader on their own terms. As stated earlier, Simon felt a responsibility to let the detectives speak for themselves no matter how unsavory their words might seem at times. In the move from page to screen, and from fact to fiction, any documentary elements that remained in the television series were the product of carefully considered decisions about visual and narrative style. The style of the series was part of a strategy for making the series more dynamic in light of its focus on character over action. But the style should also be seen as a response to the claims of realism offered up by reality shows like *COPS*: a commentary on the construction of images and on the problems of representing reality.

In this section, I will offer examples from two specific episodes in order to illustrate the ways in which the series used a self-reflexive style to call attention to itself as a constructed text. Both examples come from the fifth season of the series and feature the character of J.H. Brody (Max Perlich), the squad's videographer. Brody was introduced as a character during the fourth season and became something of a stand-in for David Simon – a mostly unwelcome presence in the squad room and at crime scenes, recording the detectives as they worked. Brody's primary function in the series was to heighten the self-reflexive elements of the storytelling that had been diminished as the network encouraged the producers to clean up the look of the series.

"Bad Medicine"

The first example comes from the episode "Bad Medicine," which featured three primary storylines. The central story concerns Det. Lewis and Det. Stivers (from the narcotics unit) as they track down a dealer selling bad heroin. The second major storyline concerns accusations of corruption against Lewis's partner, Det. Kellerman. Finally, the third storyline involves Brody's search for a new place to stay after Det. Munch kicks him out of the apartment they had been sharing. Brody ends up at Det. Bayliss's apartment. I will focus on one particular scene that juxtaposes the first and third storylines through cross-cutting and, in doing so, opens up a space to consider the construction of the scene itself.

The scene opens on the Baltimore skyline and we hear violins playing feverishly on the soundtrack. The violins actually belong to the soundtrack of a "Mighty Mouse" cartoon that Bayliss is enjoying by himself. His viewing is interrupted by a knock on the door, which turns out to be Brody. The lighting in this setting is relatively flat, approximating natural light, and the music on the soundtrack is diegetic. The scene then shifts abruptly to an exterior shot of Lewis and Stivers, along with several uniformed officers, moving in to begin their raid. They knock down the door of the dealer's house and enter into complete darkness: their flashlights the only source of light. Unlike the previous setting, the lighting here is fairly expressive – darkness punctuated by single beams – and the music on the soundtrack is non-diegetic: pounding drums. We are then taken back to Bayliss' apartment where Brody, seeing that "Mighty Mouse" is on television, suggests that they might watch a Frederick Wiseman retrospective on cable instead. The drug raid again interrupts and now the police (still in darkness) are wrestling various members of the house to the ground. It is absolute mayhem. Back at Bayliss' apartment, Brody explains that *Titicut Follies* will be showing, and proceeds to explain to Bayliss that the film is about a mental institution. The mayhem continues as the police move slowly into a room where the drugs are made. They find a light switch, which reveals a refrigerator. Upon opening it, they find the drugs they were looking for. Back at the apartment again, Brody continues to explain Wiseman's unique approach to Bayliss:

BRODY: He doesn't even put film in his camera for the first few weeks at work. Only when his subjects forget that the camera is there does he really begin to capture reality.
BAYLISS: He doesn't put film in his camera?
BRODY: It's brilliant when you think about it.
BAYLISS: (Thinking about it) Mighty Mouse. (He then turns up the volume.)

The camera now turns to the television in Bayliss' apartment to show the last scene of the Mighty Mouse cartoon, which involves a house with several large cannons shooting from the roof while a bomb descends upon it. The bomb finally hits the house and it explodes.

There are three different ways that this scene serves as a site of critical commentary on the police genre. On the narrative level, there is a tension created through the inter-cutting of Bayliss' apartment with the drug bust, which increases during the clip as the two segments are cut shorter and shorter. The tension that is created here, particularly between the two narrative threads, is one that is central to the police genre, especially since the 1970s: the tension between public and private life. In particular, this sequence highlights the disruptions of domestic spaces that are often made public through criminal investigation.

While the earliest manifestations of the genre, such as *Dragnet*, indicated that the officers had no life outside of the stationhouse, the arc of the genre over time has steadily worked to expand the representation of cops to include personal relationships and even some semblance of home life, as troubled as it might be. A telling example of this latter tendency can be witnessed most directly through Frank Furillo's negotiation of these two realms in *Hill Street Blues*. Furillo's former private life, embodied by his ex-wife, continually intrudes on his professional (public) life, while the urgencies of his job as a police Captain forever threaten and often defer his current relationship with Joyce Davenport, a public defender.

Homicide throws the tension between public and private space into bold relief, isolating and examining it. By situating two such utterly different scenes in dialogue with one another through abrupt cross-cutting techniques, the segment creates a tension that goes beyond disruptions in the personal lives of the cops and suggests something more structural: a sense of order and chaos not simply out of balance in the person of the cop, but entangled in the very fabric of social life. Through the technique of parallel editing, the sense of privacy and wellbeing so often associated with the domestic space is placed in perilous proximity to the chaos of the world outside. Even the television program in Bayliss' apartment – cartoon violence in the form of "Mighty Mouse" – offers up a contrast to the tenuous tranquility of the domestic space.

In fact, within the brief space of this scene, there are three "invasions" of the domestic space enacted. It is significant that Bayliss is watching (and enjoying) Mighty Mouse – a restorer of order – when Brody arrives. First, Brody has invaded Bayliss' home (especially by going into his bedroom) and continues to disrupt the space by questioning Bayliss' choice of programming. The second invasion of a domestic space (albeit a questionable space in terms of domesticity) occurs in the parallel scene when the officers break down the door to the house and begin arresting suspects. The chaos of this invasion is heightened by the haphazard flashlight beams that puncture the darkness. This chaos is resolved when the detectives locate the drugs – a moment significantly marked by an overhead light being switched on and its light filling the room, offering a counterpoint to the restless flashlights. Finally, back in Bayliss' apartment, once a show has been decided upon, the cartoon itself comes to a close with the outright destruction of a house to the accompaniment of the last few dissonant and unresolved notes of the feverish violin music.

Even though both scenarios have established a certain closure, the order inherent in those closures is left ambiguous (they located the drugs, but now what?); perhaps it has even been denied (the house in the cartoon has been destroyed violently). The lack of a clear resolution and satisfying closure for each episode has been a component of the genre since *Hill Street Blues* in the early 1980s dared to keep as many as seven different storylines active over several episodes without any definitive resolutions. But the important point to be made about *Homicide*'s approach in the sequence described above is that it exists primarily as a commentary on the nature of resolution itself: that it is a construction of the genre, as natural as the antics of a cartoon super-hero.

Finally, there is also a stylistic tension at play in this scene. This tension is played out between the two settings. The scene in Bayliss's apartment is shot in long takes, with natural sound and a handheld camera. The camera angles and movements are reminiscent of the *vérité* style of the dispassionate observer: simply following the action as it unfolds. The drug raid, however, is much more dynamic in its construction: pulsing non-diegetic music, overlapping editing as the baton hits the door, a quick editing pace, and dislocation of space during the raid. This distinction is important in that it neatly encapsulates the primary stylistic tensions of the series as a whole: the tension between the need to provide an entertaining, dynamic story, and the desire to use that dynamism to comment on the high degree of construction and mediation in these images.

The significance of this tension for the purposes of this argument lies in the police genre's dual role as both popular entertainment and a site where specific codes of realism lend a certain weight and credibility to representations by creating an air of authenticity in the images themselves. While the specter of authenticity is surely dubious, even in traditional documentary forms, *Homicide* confronts and highlights the ways in which the genre typically constructs authenticity through the codes of realism. The important point to keep in mind is that *Homicide* does this work from *inside* the genre. The critical function of the scene, in fact, can be located in its self-reflexivity.

For instance, Frederick Wiseman (the subject of discussion in the scene) often employed the tools of parallel editing to juxtapose actions and create a meaning that was greater than those located in the individual sequences. By bringing the segments into dialog with one another, a second order meaning is established. The fact that Brody and Bayliss are discussing Wiseman while the show juxtaposes the two disparate scenes calls attention to the style of the series itself. Just as Wiseman critiques the very institutions he was allowed to observe, *Homicide* offers a critique of the police genre as a kind of cultural institution.

Clearly, this brief scene stands as a moment of critical reflection and commentary – a moment in which we can witness a television text looking not only at itself but at its existence as part of a larger cultural construct: the police genre. In other words, it offers up a critical examination of its own conventions as well as the conventions of documentary filmmaking. My point here is that this examination is not simply an exercise in pastiche, but rather an effort to complicate the assumptions that surround all representational conventions and, in doing so, perhaps stretch the boundaries of the genre, especially with regard to realism and verisimilitude.

Verisimilitude, of course, is something towards which the police genre has always made overt gestures, though in different ways at different times. While *Dragnet*'s realism relied heavily on the mundane, exacting nature of police work along with the use of real cases "pulled from the files of the LAPD," shows like *The Naked City* and *M Squad* made their case for authenticity by shooting on location in New York and Chicago, respectively. With all this in mind, it is possible to think about *Homicide*'s style not as realistic, *per se*, but as incorporating a host of strategies handed down from its own generic past. Like *Dragnet*, the chases are few and gunplay is rare; it's the work that matters. As with *The Naked City*, it is shot exclusively on location, and like *Hill Street Blues* it owes its greatest stylistic debt to documentary filmmakers like Frederick Wiseman and Alan and Susan Raymond. More than anything else in the show, it is arguably its connection to documentary that provides the most provocative space for thinking about *Homicide*'s critical function, since so much of what the show has to say seems tied to this stylistic identity and practice.

In the next section I will examine the single most self-reflexive episode produced for the series in order to highlight the degree to which *Homicide* embraced the tensions between reality and its representation.

"The Documentary"

Perhaps the most telling and extensive example of the critical function of *Homicide*'s style can be found in another episode from its fifth season, "The Documentary". The episode takes place during one slow New Year's Eve inside the homicide squad room. In order to pass the time waiting for the "ball to drop" (and the phones to start ringing) the squad watches a documentary about themselves, produced by Brody, the crime-scene videographer. From the moment Brody places the tape in the VCR, the world of the show and that of the documentary become increasingly difficult to distinguish. The opening credits for the actual show roll as if they were Brody's invention, part of his creative effort, drawing a link between that which is entirely fictional (*Homicide*) and that which is purported to be real (Brody's documentary). Throughout the episode, the squad, forever contending Brody's efforts and motives, engage in a running commentary, critiquing Brody's stylistic choices ("Ooh, montage, my favorite!"), and challenging the propriety of his efforts ("You can't show that. That's private."). Although Brody's documentary can in no way be mistaken by the viewers of *Homicide* as something real, within the space of this episode, it must be considered as a document. The fictional characters we have come to know over the weeks and years suddenly take on the air of real people as they sit alongside us, watching their lives and work undergo videographic dissection.

The bulk of Brody's documentary follows the arc of one particular case that involves the murder of Llewellyn Kilduff and his wife by their next-door neighbor, Bennett Jackson (played by Melvin van Peebles). Upon arriving at the scene, Detectives Pembleton and Bayliss discover that Mr. Jackson has already admitted to the murders and is willing to go with them to the police station and sign papers to that effect. Pembleton is pleased. His job has been made that much easier. The facts are clear: he knows the answers to who, what, when, where, and how. Bayliss, however, is not so satisfied. The following exchange takes place in the car driving back to the squad room. Our vantage point is Brody's: from the back seat, as if looking through the camera ourselves:

BAYLISS: You might think this case is a slam dunk...
PEMBLETON: Our case is being writ in black even as we speak.
BAYLISS: You'd be wrong.
PEMBLETON: No, you'd be right! We've got the shooter, we've got his gun, we've got beaucoup eyewitnesses, and the man's given it up.
BAYLISS: (whispering to the camera) We need the *why*.
PEMBLETON: No, *you* need the why. I don't need to know any more about the man or his problems than this: he shot his neighbors, then waited on his swing for the police to arrive so that he might surrender his freedom. Mr. Jackson has been so helpful and so efficient that to ask for more would be ungracious.
BAYLISS: Come on, Frank! One neighbor murders another and you don't want to know what that means?
PEMBLETON: I know exactly what that means: ten hours overtime and if Mr. Jackson would be kind enough to take this to trial, another twenty hours court pay.

In a very short space of time, this scene lays out the professional tension between the two detectives as well as subtly alluding to the show's position within the genre. Pembleton's words could come almost as easily from the lips of a stoic Joe Friday or his partner, Frank Smith: "We're just interested in the facts." But that kind of certainty can no longer hold. These cops have lives and emotions. They are not the automatons of bureaucracy that once walked these streets; they are curious and eager to make sense of their surroundings and learn something about themselves in the process. Motive is usually only a means to an end. Once the motive has been determined, the criminal can be apprehended and the detectives can move on to the next case. Bayliss, however, is haunted by his need to understand the underlying motive, and he is not alone. In fact, it is a defining characteristic of the series that a name in black on the board does not necessarily relieve the personal resonances, the feelings of responsibility, the unresolved residue of moral misgiving that often linger long after a case has been put down. In the case of Mr. Jackson, Bayliss cannot be satisfied with the bare facts alone; he seeks to understand the situation that brought about the murder.

The desire to know eventually leads Bayliss to discover that Mr. Jackson, a mortician, has been bringing dressed up corpses to his home in order to assuage the loneliness he has felt since his wife entered a nursing home. His neighbors, having found him out, paid with their lives for their knowledge. This knowledge is shocking enough on its own. Bayliss and Pembleton can hardly disguise their disgust at a man who would stoop to such morbid depths – so low as to have his picture taken with the corpses. But this is not mere necrophilic fantasy; it is a psychosis borne of loneliness, an emotion that Bayliss can all too easily understand. Throughout the course of the series, Bayliss has been depicted as a man in search of meaningful companionship, both romantic and collegial. He begins the series as an outsider among the homicide detectives who has never succeeded in maintaining a sustained romantic relationship. Bayliss's search for the answer of *why* Mr. Jackson would kill his neighbors stretches beyond the boundaries of a murder mystery or procedural drama (the limits that Pembleton places on the case); it becomes a personal mystery as well.

The lesson here for the police genre is: "Be careful what you wish to know, for you just might find out." Joe Friday and his colleagues could hide safely behind the bare facts, never confronting the possibility that they might suffer the same psychological malaise found in their suspects. As the private lives of the police have taken on more narrative weight, a new kind of dangerous knowledge emerges. Humanizing the police force, it turns out, is a double-edged sword. Where once the lines could be drawn with certainty between those who threatened the social order and those who maintained it, *Homicide* demonstrates just how easily this line can be shifted, even erased.

As a way of underscoring this humanizing tendency, Brody's documentary contains several moments that expose the extremely private moments of the detectives' lives. These scenes play off and against the historical trajectory of the genre, but also highlight the ways in which this episode simultaneously enhances the series' realism and comments on the constructed nature of that reality. This tendency is perhaps best represented by the following scene, which begins with the title "Off Duty."

Detectives Lewis and Kellerman approach the door to the Waterfront Bar, derisively acknowledging Brody's presence ("Get a life, Brody"). Brody follows them through the door as they approach the bar, sit down, and order a beer. At this point, the exact same scene repeats, eliciting a critical response from the squad room audience:

BAYLISS: Didn't we just see this?
COX: Yeah, Brody. Obviously you screwed up.
BRODY: It's a choice. It's a cinematic statement.
COX: Yeah? It looks like a mistake to me.
PEMBLETON: I like it. It speaks to the repetitive and essentially meaningless nature of police work.
COX: This I could do without.
BAYLISS: Wow, this is really dramatic stuff.
KELLERMAN: A glass of beer.
GIARDELLO: Yes, Brody, the whole thing needs more action.

On one hand, the scene is clearly referencing the show itself. Barry Levinson always envisioned a series in which style would stand in for action; instead of car chases and gunfights there would be jump-cuts, triple takes, and a dynamic camera. On another level, though, the scene is also a sly comment on the trajectory the genre has taken over the years: getting more and more involved in the intricate details of the cops' personal lives. Where we could once take pleasure in watching Joe Friday grinding through the facts of the case (and nothing else), there is now the danger that something as utterly mundane and far removed from police work as the pouring of a glass of beer might become a central component of the narrative. Stylistically, the scene brings ideas about the construction and manipulation of reality to the surface. As Dr. Cox's comment suggests, we have certain expectations about what connotes reality and when those are not met we feel that somehow a mistake has been made. Though we never get to hear Brody explain what kind of "cinematic statement" he intended here, we might imagine, in addition to Pembleton's reading, the point to be made is that there is no such thing as reality on the screen. Once actions have been turned into electronic images, they enter the realm of the imaginary, of signs and language and, by extension, arbitrariness and "cinematic fakery."

Additionally, turning back to Pembleton's reading of the scene, we can also see that the series is a comment on the nature of police work itself and the mythic character of the detective-hero. In the world of *Homicide*, even the presence of the detective does not guarantee explanations. While they are often successful in solving their individual cases, they tread on dangerous ground if they begin to search for deeper meanings. On the one hand, to ask the question of why a murder was committed opens the detective up to the kinds of revelations that may ultimately turn inward. Tim Bayliss's pursuit of the underlying motive behind Mr. Jackson's murder of his neighbors led not to any great cathartic releases – no cleansing epiphanies of the soul – but to a disconcerting acknowledgement that the line between depravity and virtue is a thin one indeed. To understand another human being's motives for malice is to acknowledge one's own dangerous potential. When Bayliss discovers the truth about Mr. Jackson, he is compelled to empathize with his loneliness, if not his unsavory solution.

As these two examples illustrate, the stylistic choices that the producers of *Homicide* made were motivated by a range of factors. First, and most obviously, the style was driven by the need to set the series apart from other dramas in the prime-time lineup on broadcast television. But commerce only goes so far as an explanation, and certainly does not determine the range of meanings available to any given viewer. I argue that the extreme self-reflexivity of the series (embodied in the examples above and manifesting perhaps less overtly across the majority of the series episodes) establishes *Homicide* as something of a critical text: reflecting on the very nature of reality and the realist aesthetic to which the police drama on U.S. television has so frequently clung. By calling attention to the aesthetic choices made by producers, the series opens up a space for critical reflection on the ideological stakes of representation itself.

Homicide and the Symbolic Construction of Community

In addition to its willingness to reflect on its own aesthetic genealogy, *Homicide* also opens up a related space for critical reflection on the notion of community. Police dramas may, in fact, be understood at the most basic level as dramas about the protection and maintenance of communities. But how those communities are defined is rarely a source of speculation or reflection. One of the key arguments of this chapter is that such speculation and reflection is a central element of *Homicide*'s dramatic world. Indeed, in keeping with the larger argument of this book, *all* police dramas participate in this dialog. *Homicide* merely makes the conversation manifest.

In the opening sequence to a third season episode ("Colors"), Detectives Bolander and Pembleton have a seemingly tangential conversation while on their way to a murder scene:

BOLANDER: Nothing is real.
PEMBLETON: What do you mean: nothing is real?
BOLANDER: There is no reality.
PEMBLETON: Really...
BOLANDER: Take the color green. You see green. I see green. We call it "green" because as a society we have agreed that this thing, this color, is green. We think we're having the shared experience of green, but how do we know? Maybe my green is actually greener than your green.
PEMBLETON: You mean maybe my green is red?
BOLANDER: Maybe. Take colorblind people, they carry with them a stigma...
PEMBLETON: Astigmatism.
BOLANDER: Because they don't see what the rest of us see as green. But maybe, just maybe, their perception is correct. Maybe a colorblind person is actually seeing pure green, the *real green*.

Bolander's incorrect assumption that the inability to see green *stigmatizes* colorblind people is more than just a humorous play on words. His mistake highlights an important part of culture and the concept of community. One way to define culture is as a network of shared meanings and interpretations that are not always completely in agreement and are subject to unequal power relationships, but are at least overlapping in significant ways. And these meanings or interpretations are constituted symbolically: they are products of *representations*. Culture also involves the *process* of communication – of negotiating these potentially different meanings with one another. The less able we are to share meanings with one another, the more likely we are to see ourselves as separate from others: even to the point of stigmatizing these differences in others.

Similar to the manner in which *Homicide* provides a space for critical reflection on the realist aesthetic that has been so central to the police drama historically, the series uses its overt televisual style to explore the very notion of community, calling attention in particular to the symbolic construction of communities. By symbolic construction, I mean what Gerard Delanty (pointing to the work of Anthony Cohen) has suggested: that "community exists ultimately in the symbolic order rather than in an empirical reality; it is a form of consciousness or awareness of reality" (2003: 46). This interpretation of community moves away from more traditional notions of the concept that are based in either geographic or demographic proximity. In Cohen's view, community:

> is a largely mental construct, whose "objective" manifestations in locality or ethnicity give it credibility. It is highly symbolized, with the consequence that its members can invest it with their selves. Its character is sufficiently malleable that it can accommodate all of its members' selves without them feeling their individuality to be overly compromised. Indeed, the gloss of commonality which it paints over its diverse components gives to each of them an additional referent for their identities.
>
> *(1985: 109)*

Thus, community is not a "compelling moral structure that determines behavior but is a resource from which people may draw" (Delanty 2003: 47). But, of course, those grounded, empirical facts of ethnicity and locality that always adhere to notions of community remain vital even if one insists, as Cohen does, that a sense of 'community' is largely a matter of locating meaning in a relatively open system of shared 'cultural maps' rather than abiding by any monolithic and reductive sense of belonging. As Cohen states: 'people assert community, whether in the form of ethnicity or of locality, when they recognize in it the most adequate medium for the expression of their whole selves'. (Cohen 1985: 108).

But Cohen does not simply or naïvely ignore the fact that, despite its largely symbolic nature, community is still a concept that matters greatly to our notions of belonging (and *not* belonging). In fact, Cohen argues that the word "community" implies both similarity and difference. "The word thus expresses a *relational* idea: the opposition of one community to others or to other social entities." Communities, according to this definition, rely on *boundaries,* which "are marked because communities interact in some way or other with entities from which they are, or wish to be, distinguished." These boundaries may be geographic, racial, linguistic, or religious (to name a few), but whatever the form of the boundary, it must have at least some shared meaning for members of the community. Boundaries "encapsulate the identity of the community" and, by extension, the identities of the individuals that make up the community (1985: 12).

The series is, of course, a product grounded in a very particular time and place (Baltimore in the 1990s). But within this time and place, the series explores the multiplicity of identities, interactions, and meanings that are available to members of this community. It allows us to consider smaller communities, identified more by race, class, gender, and occupation than by representation within the political state (i.e. the citizens of Baltimore). Specifically, the series often contemplates the ways in which we construct our communities through symbolic engagements with tradition, the shared past, shared experience of the present, convergent as well as divergent political aims, interests, and experiences. By highlighting the construction of representations (as discussed earlier) the series is also able to highlight the symbolic construction of the very idea of community and to comment critically on the ways that the police genre has relied on traditional notions of community driven mostly by a focus on legal rights and responsibilities.

In what follows, I will offer an analyses of three particular episodes of *Homicide* that illustrate different ways in which the series addresses the symbolic construction of communities. At the heart of my analyses are the formal, stylistic approaches that the series' producers take to the issue. Highly self-conscious stylistic devices (such as color filters) as well as more standard techniques (such as manipulating depth of field) provide a layer of commentary about how communities are structured and maintained. Each case study highlights a different aspect of community: the tension between stasis and change; the community as the provider of a sense of place and belonging; and the community as defined by a set of meaningful codes or signs that establish boundaries (even if only temporarily). Cutting across each of these ideas are the geographic boundaries that mark urban communities; but those physical markers are only a starting point. Underneath the urban geography is another kind of cultural geography based on race, class, and gender that make the concept of community much richer, more complex, and more contradictory than any map can show.

"Every Mother's Son": The More Things Change...

As its title suggests, "Every Mother's Son" is a story, not about individuals, but about relationships and connections. It is also a story about stasis and change within communities. The episode highlights not only the changing physical geography of Baltimore (neighborhoods and business districts) but also the cultural geography of race, class, and gender, especially with regard to questions of change and preservation. Who has the ability to change their circumstances? What are the political stakes in preserving the past?

The case of "Every Mother's Son," involves a 14-year-old child, Darrel Nawls, who has been shot to death by another 14-year-old, Ronny Sayers, in a case of mistaken identity. What is most compelling about this particular episode is the fact that the narrative focuses less on the detectives and the case at hand, and more directly on the relationship between the boys' mothers. In fact, the pattern of murder, pursuit, apprehension, interrogation, and confession, which typically constitutes almost the entire narrative arc of a traditional police story, is taken care of within the first half of the episode. In short, there really is no investigation of the *crime* to speak of. There is never any real doubt as to who shot Darrel and, once arrested, Ronny confesses immediately (mostly because he sees nothing wrong with what he has done). What is left, then, is another type of investigation: an investigation of the kinds of circumstances that lead to this type of violence.

In a particularly important and illuminating sequence, the episode compresses three scenes (notifying Darrel's mother, her identification of Darrel at the morgue, and the detectives' attempt to arrest Ronny at home) into the space of just over two minutes. A song on the soundtrack accompanies the entire sequence, the lyrics of which comment on the images and underscore the parallels between the two boys as well as their mothers. As the sequence moves from the identification of Darrel to the pursuit of Ronny, the singer intones: "See this child, twice stolen from me." Not only does the image of a stolen child resonate strongly with Darrel's death, but it is also designed to represent Ronny. Darrel has been "stolen" by death and Ronny will be taken by the police. Beyond these very apparent plot-level correlations, though, the idea of *twice* stolen suggests that they were each somehow already gone – their innocence taken by a life in which they are forced to live amongst violence and suffering each day. In this way, the image resonates back to the mothers, as well as their other sons, the younger brothers of Darrel and Ronny. "Twice stolen" simultaneously resonates with the two mothers who will each lose their eldest son, and also signifies the threat of losing their second children in similar fashion to the same deadly momentum of the inner city.

The two women and their sons seem caught in a cycle of violence and relative poverty, stuck in deteriorating neighborhoods that lack strong male role models for their sons. This predicament is captured in a long conversation that the women have while sitting in the "fishbowl" – the waiting room in the homicide unit:

MRS. SAYERS: Where are all the men these days? That's what I want to know.
MRS. NAWLS: In my neighborhood...there's Gerald. He's twenty-six. [Laughs] Then there's Marcus. Marcus has got to be eighty years-old. But that's just my neighborhood.
MRS. SAYERS: There's this one guy...he might be in his thirties. I haven't seen him in a while. I don't know...maybe something happened to him.

As their conversation continues, they begin to talk about their kids (at this point they are not aware of their relationship to one another through their oldest sons):

MRS NAWLS: I've been to three funerals for Darrel's friends this year. Seems like that's how we have our socials now.
MRS. SAYERS: I'm gonna get my boys out of here. Every place I can think of costs money. But I'm thinking about Canada. Maybe they don't kill each other so much up there.

MRS. NAWLS: Canada, huh?
MRS. SAYERS: Girl, I don't know nothin' about it except it's supposed to be a whole bunch of snow and cold. Maybe the snow keeps a lid on things – d'you think?

There are several issues at play in this brief exchange. First, the women discuss the shortcomings of their neighborhoods – the foundations of their physical community – especially the lack of acceptable men, either as role models for their sons or, possibly, husbands or boyfriends for themselves. Given the deteriorating state of their neighborhoods, Mrs. Sayers imagines leaving for a safer place, even if it means completely uprooting herself and her kids. Importantly, she alludes to the fact that her mobility is severely limited by finances. In a relatively simple exchange, the writers of this scene confront issues of race (the loss of black men to prison or violence), class (limited finances for relocation), and gender (the pressures of raising children as a single mother). Each of these issues underscores the shared circumstances of the two women. The important point is that these circumstances are both material (financial and geographic) and symbolic. They are symbolic because they have meaning for the women and this meaning overlaps enough that they can commiserate and connect with one another.

In addition to the shared ground that the conversation uncovers, the style in which the scene is blocked and shot emphasizes the connection between the two women. The scene takes place in a relatively small space and the women are sitting on a bench, facing the same direction. The scene begins with the women a few seats apart, and the shot is composed with a wide angle lens in order to lengthen the depth of field and keep everything in focus. As the conversation progresses Mrs. Sayers moves closer to Mrs. Nawls. With her shift in position, the composition of the shot also changes slightly. More importantly, the depth of field narrows significantly, indicating that the camera is now equipped with a longer lens. Shortening the depth of field compresses the perceived space between the two characters. Furthermore, a shift in shot composition from a medium close-up to a close-up, combined with the shortened depth of field, has the effect of layering the two women's faces onto one another. Like the lyrics of the song in the earlier example, these camera techniques work together with the dialogue to accentuate the commonalities between the two women.

While the episode gives the connections between the two women a great deal of attention, it also focuses on community divisions as well – both physical and social. This issue of community division is captured in the juxtaposition of two scenes that, on the surface, appear to have nothing in common. The first scene involves the police capturing Ronny. Bayliss and Pembleton, along with a team of uniformed officers armed with shotguns, corner Ronny in a back alley. Ronny pulls a pistol from his belt and begins waving it around at the police and eventually holds it to his head. As the camera swings between Ronny and the detectives, ramping up the tension, it captures the texture of the setting as well: graffiti on buildings, burning trashcans, weeds growing up through the cement, and discarded tires. Eventually, the detectives close in on Ronny and Bayliss tackles him onto a pile of tires. The narrative then shifts to the inside of the Waterfront Bar, which is owned by three of the detectives, where an interior decorator (Howie Mandel) is providing an assessment of the kind of work that needs to be done. When he goes to look at the bathroom, a woman from the "Baltimore Landmark Preservation Society" enters the bar and tells Munch and Lewis (two of the three owners) that she is there to "protect the integrity of the property." As she says this, the decorator emerges from the bathroom and sighs: "We definitely gotta do something about that bathroom. It's like from the Civil war or something. We should just rip everything out and start from scratch." The woman lowers her gaze at him and shakes her head. She then informs Munch and Lewis that the building cannot be altered in any way because of its connection to American history: it once served as a bathroom stop for George Washington on his way home from a party.

On the surface, the scene in the bar seems to serve two functions. First, it continues a minor plotline (buying the bar) that cuts across several seasons. Second, it provides some comic relief after a tense scene. But the fact that these scenes are juxtaposed with one another creates another level of meaning that connects to the scenes discussed earlier. The fact that the two settings contrast with one another – an abandoned lot in a deteriorating neighborhood versus a historic landmark in a gentrified section of the city – highlights the real physical dislocation between the two.[3] More importantly, this distinction points to the social and cultural boundaries in the city. Ronny is literally cornered in the vacant lot and is eventually thrust onto the heap of discarded tires: symbols of freedom and mobility reduced to unwanted trash. Meanwhile, the bar stands as a testament to a community in flux; it has new owners, is located in a revived section of the city, and is said to be worth preserving.

The tension between stasis and change, between renewal and preservation, which is at the heart of community identity, is a key to the episode. If communities are symbolically constructed, it is important to understand how members of a community make sense of their place within the community. What is meaningful? To whom is it meaningful? As Anthony Cohen states:

> The manner in which the past is invoked is strongly indicative of the kinds of circumstance which makes such a "past-reference" salient. It is a selective construction of the past which resonates with contemporary influences. Sometimes this kind of folk history resembles myth, or meta-history, in the sense which Malinowski gave to the word: a "charter" for contemporary action whose legitimacy derives from its very association with the cultural past. Myth confers "rightness" on a course of action by extending to it the sanctity which enshrouds tradition and lore.
>
> *(1985: 99)*

The city officials have *chosen* to preserve the building. Stasis is history, a meaningful monument to the past. Change is possible, just undesirable in this case. Mrs. Sayers and Mrs. Nawls, however, do not have the luxury of choice. For them, stasis is inevitable – it means decay and tragedy – and substantive change is out of their reach. It seems safe to say that the historic building as a symbol of the community doesn't mean very much to these women, or their sons.

"Narcissus": A Sense of Place

The kinds of social and physical dislocation that I pointed to in "Every Mother's Son" are also central to "Narcissus," which aired in the series' fifth season. This episode deals with community as providing a sense of place – a sense of belonging. It also deals with the physical, cultural, and social boundaries that we erect in order to maintain our sense of place, our sense of our own community, and investigates how media representations, and our consumption of them, form that sense of place.

This episode deals with a standoff between the Baltimore police and a community group called the African Revival Movement. One member of the movement has murdered another and is holed up in the movement's headquarters, a large building that houses the administration of the movement as well as several families. The African Revival Movement is lead by Burundi Robinson, a former police officer turned social and political activist. Robinson is seen as a hero in the neighborhood, providing opportunities and a sense of purpose for young, poor, black men and women. He is also popular with the press, providing them with access to his shining example of community building. At the same time, Robinson is a disgraced former police officer, having left the force in light of a drug scandal involving him and a fellow officer, now the Chief of Police. He had since used his knowledge of the Chief's checkered past opportunistically in order to gain personal and political favors.

In order to establish themselves as a community of their own, in solidarity with their African heritage, and in order to distance themselves from the American community they see as hopelessly racist, the members of the African Revival Movement have taken the names of African nations: Burundi Robinson, Kenya Merchant, Malawi Joseph, and Benin Crown. This act of defiance symbolizes the fact that these individuals refuse their identities as black Americans. This symbolic refusal, coupled with the material monument of their communal living space, places them outside the mainstream social structure of America. In essence, what they have refused, both symbolically and materially, is the political neglect and socially subordinate position that often goes along with being a member of a minority group in a racist society.

This act of constructing both a symbolic and a physical boundary is designed to create a sense of place among the members of the African Revival Movement. As Doreen Massey points out, "a 'place' is formed out of the particular set of social relations which interact at a particular location" (1992: 12). Thus, while *space* is a physical phenomenon, *place* is a social one. Communities, then, can be marked geographically as *spaces*, but they are also marked, more centrally, by the social relationships that occur within them. Communities are *places*. But Massey is careful to point out that *place* in this sense does not mean an isolated community. Instead, she insists that the identity of a *place* "derives, in large part, precisely from the specificity of its interactions with 'the outside'" (1992: 13). It is in this sense that the episode in question takes on the most weight as a discourse about community. Having barricaded themselves in the compound, the members of the movement foreclose any more involvement with "the outside" and thus jeopardize their ability to maintain a meaningful identity. This problem is made tragically manifest in the closing sequence of the episode.

As the standoff reaches into the night, the word comes down that there has been no sound coming from inside the compound for several hours. Sensing that something has gone terribly wrong, the police storm the headquarters. Their flashlights puncture the darkness, nervously scanning the threatening space inside and occasionally sweeping across fearsome tribal masks hanging from the walls. Stairwells and hallways reveal nothing but an uncomfortable emptiness. A beam of light settles on a closed door and the police quickly break through. There is a sudden and complete silence as the probing beams reveal a small room filled with dead bodies stretched out on the floor; Robinson is found sitting up in a chair. It is a mass suicide.

Outside again, as the police remove guns and bottles of poison from the compound, Lt. Giardello stops before a crowd of reporters as he is asked for an explanation of the tragedy. "Why? Why did he do it?" asks one reporter. "I don't know," says Giardello, "but I can tell you one thing: there's going to be Hell to pay." With that he moves through the crowd.

The scene then shifts abruptly out of the narrative world of the crime scene and into an anonymous suburban living room where a couple, heretofore unseen, sits watching an account of the event on the evening news while they eat their dinner. At the completion of the news broadcast the man picks up his remote control and switches the channel. Finding another account of the event, he quickly moves on, resting finally on what appears to be a PBS program about marine life off the coast of Maine. The camera then moves to an adjacent dining room where, from the vantage point of an empty dinner table, we continue to peer in on the couple as they go on quietly with their meal.

The abrupt jump to the suburban space is an important and illustrative moment in the episode. Without warning, the show deftly slips out of its own narrative skin and offers a commentary on the implications of what we have just seen. Tellingly, this anonymous space is introduced by a shot of a television screen on which we can see Lt. Giardello move through the crowd of reporters and then off screen (matching the action from the previous shot of Giardello at the scene). As a reporter's voice begins to describe the incident, the camera in the living room pans from the television to reveal the young couple on their couch, eating takeout Chinese food while watching the newscast. The point being made in this brief sequence is, in many ways, the same point made by David Morley and Kevin Robins: "The screen is a powerful metaphor for our times: it symbolizes how we exist in the world, our contradictory condition of engagement and disengagement" (1995: 141). Simultaneously engaged (in watching) and disengaged (far removed from any active participation in the event), the young suburbanites are consummate consumers. From the comfort of their living room sofa, the couple is able to devour, like so much Chinese food, the tidy packages of information about the street just outside. It is a way of confronting the 'Other' – of harnessing difference through an act of consumption.

Having consumed the necessary information (indicated by the conclusion of the news report), the audience is free to move on. As the television image settles into the PBS-like program about marine life, the audio from their television intones: "From Kittery to Kennebunk, the coast of southern Maine is home to a unique variety of marine life that thrives in the cooler waters away from the gulf stream current." To be sure, it is difficult to imagine anything further from the troubles in West Baltimore than a show about marine life off the coast of southern Maine. The brief segment of audio (which also concludes the episode) serves as a counterpoint to the rage and catastrophe we have just seen in the inner city and, as such, serves to underscore the problems of unity and diversity in the United States. It is important that the "unique variety of marine life" of which the narrator speaks, thrives away from the tumult of the gulf stream current. The turbulence of the current becomes a handy metaphor for social unrest, while the "cooler waters" suggest an impossibly utopian harmony – a mirage refracted through the television tube.

The episode, like the series itself, however, offers up an ideologically complicated (perhaps even contradictory) vision of justice and community. On the one hand, the members of the African Revival Movement, by changing their names to those of African nations (Kenya, Benin, Malawi, and Burundi) have flouted the myths of American pluralism. They have asserted their difference in a most profoundly personal and symbolic way. Given the symbolic strength of white, middle-class hegemony on American television, then, it is no surprise to witness the eventual outcome of the standoff. The mass suicide becomes a logical (though deeply unsettling) conclusion to an episode that supposes there is no symbolic place in this society for African-Americans who embrace their African heritage to the detriment of an American identity – who favor the left side of the hyphen over the right. At the same time, however, the more situated community identities that accrued around the African Revival Movement point toward the kind of nationalist community identities (via the names taken by the members of the movement) that a focus on ethnic identity (over national identity) reacted against. In this case, the community that builds up around the African Revival Movement threatens to become as monolithic and repressive as the larger civic community against which it stands. As Iris Marion Young has pointed out, the very notion of community "denies and represses social differences" (1990: 227). Graham Day, reflecting on Young's perspective, points out that a community's "desire for social wholeness encourages its inhabitants to adopt 'mystical' ambitions, which submerge individuality into group identity" (2003: 208). Thus, The African Revival Movement's ambition of existing outside of the larger civic community of Baltimore leads symbolically to the very real demise of the entire group at their own hands. Just as unsettling, however, is the fact that this symbolic act of ultimate sacrifice could be so quickly packaged for consumption and reproduction. In the end, it is representation that matters. It is no mistake that the story of the sixteen dead members of the African Revival Movement beaming into that secure suburban space refers not to any individuals but only to "victims" whose names, the names of African nations, remain unspoken and unknown.

"Colors": Reading the Signs

If communities are symbolically constructed, then one's participation in a given community rests on the ability to interpret its central symbols in ways that overlap with other members. Failure to read the signs appropriately could result in individuals being ostracized or even killed. Moments of extreme violence, in fact, often cause us to question our allegiance to certain symbols, or at least debate their meanings. It is this kind of reflection that the episode "Colors" (season three) asks us to undertake, especially with regard to how notions of nationalism, ethnicity, and race affect our sense of community.

In this episode, a young high school student, Hikmet Gersel, on exchange from Turkey, is accidentally shot and killed by a white suburbanite in Baltimore, Jim Bayliss (cousin of homicide detective Tim Bayliss). The incident has occurred as the result of a misunderstanding in which the boy (intoxicated and made up like a member of the rock group, Kiss) arrives at the Bayliss's door, thinking that he was actually going to a party with friends. The sight of the boy in his ghastly makeup startles the homeowners and the situation escalates to deadly heights, ending with Jim Bayliss shooting and killing Hikmet. In attempting to determine whether or not the shooting was an act of self-defense, Det. Frank Pembleton (Tim Bayliss's partner on the squad) and Det. Bolander interview Jim, along with his wife and a friend of the victim who saw the incident from a distance. All three recollections tell the same general story with slightly different inflections of various details, not the least of which is whether Hikmet had his hands raised in mock surrender (thus throwing any sense of danger or threat into serious or at least reasonable doubt).

The three eyewitness accounts are presented to the viewer within the first half of the episode. Each retelling bears a general resemblance to the others, though individual details change from story to story. The most striking thing about these accounts is the style of presentation. The producers added tinting to the images in the separate narratives. Jim's account of the confrontation is tinted red, his wife's is tinted blue, and Hikmet's companion's is green. On the one hand, this stylistic move simply accentuates the *Rashomon*-like quality of the episode. On the other hand, the colors refer back to the earlier discussion between Pembleton and Bolander about the nature of reality. The facts of the story may differ, but they are beside the point, ultimately. What matters is how the other members of the community *interpret* the event.

Interpretation has been the problem all along in this episode. Inability to read the signs of the encounter has caused a simple mistake to escalate into a murderous confrontation. First and most simply, Hikmet and his friend have literally misread the directions to the correct neighborhood (they were on their way to Hickory *Lane*; Jim lives on Hickory *Avenue*). Next, both Jim and his wife misread Hikmet's attire (leather jacket, wild hair, and makeup) and his behavior (drunken) as a threat. They literally don't know what to make of him, though they assume he means them harm. Furthermore, Jim literally cannot understand Hikmet's broken English. Finally, Hikmet misreads Jim's gun as a game. This series of misunderstandings, based as they are on serious cultural differences between the two primary participants, has tragic consequences. These facts are important for the police investigation, and there is very little room for interpretation within the actions themselves. The more important question is: what do these actions mean? It is at this level of interpretation that the episode engages with the issue of community through the lens of nationalism and racism.

Though Jim is generally a likable and sympathetic character, over the course of the episode, his sometimes not-so-subtle racism rears its head and casts serious doubt on his motives. His wife, for instance, admits that Jim sometimes loses his patience with people who don't speak English very well. Most disturbingly, as Jim takes a moment to wash Hikmet's blood from his porch he asks, "Who would have thought their guts would be the same color as ours?" What seems to be at work here is something akin to Samuel Huntington's clash of civilizations. For Huntington, as technological innovations create the perception of the world as a smaller place, "the interactions among peoples of different civilizations enhance the civilization-consciousness of people that, in turn, invigorates differences and animosities stretching or thought to stretch back deep into history" (1993: 26). These seemingly more basic and lasting affiliations of blood, kinship, and soil are made manifest as Jim wonders at the strange similarities between Hikmet's blood and his own.

One of the problems with Huntington's argument is that it is posed from a distinctly Western perspective. It is as if the norms of Western civilization should be the world's aspiration. These norms, or "national political values and beliefs" (Huntington 1981:13) comprise what is known as the "American Creed" – an amalgamation of seventeenth-century Protestantism and Enlightenment thinking which can be roughly defined as "liberty, equality, individualism, democracy, and the rule of law under a constitution" (1981: 14). For Huntington, this creed is the pillar of American national identity. In this paradigm, however, one is led to suppose that America, the land of immigrants, may actually find itself in jeopardy – in danger of splitting wide open in the face of the cultural diversity which has always defined the nation. Huntington's warning rings as mostly reactionary and implies a call for a return to a more thoroughly nationalist identity formation.

We can see a similar tension at play in a particular scene near the end of the episode. In front of the grand jury (to determine whether or not he should be charged with murder) Jim, a lawyer, delivers a moving speech on the reasoning behind his actions:

> There's a tried and true proverb: "A man's home is his castle" ... The place you sleep and eat, the place where you read or watch TV, where you play with your children or make love ... If you're lucky enough to have a home, you have the right to defend it. You have the right to make sure that your children are safe, that your wife, your husband, your long-time companion is safe. No one should ever be forced to flee from their own home. It is our ultimate sanctuary ... Do I feel good about what I did? No. I will have to live with the horror of that split second for the rest of my life ... Do I believe I did the right thing? Yes... Yes... Yes... Yes... Yes... Yes... Yes. And I know that you know that given the same circumstances you would behave exactly as I did.

Not surprisingly, this speech succeeds in swaying the grand jury to acquit Jim of murder – a decision that is met with fervent applause from the crowd in the courtroom. The heroic, property-owning individual has been vindicated by a jury of his peers.

This scene plays out quite clearly the constitutional guarantees of life, liberty, and the pursuit of happiness which Americans so lovingly embrace. It is a reality of living in the free world – an expectation of inalienable rights. What Jim's speech does so effectively is set the symbolic stage for justifying his actions in the name of fairly traditional notions of community and civic responsibility: a portrait of middle-class aspiration and innocence that is sure to resonate with his "audience." His view is nostalgic, tied to a past that his interpretation of Gersel's actions threatened. As Cohen suggests, "if the individuals refer to their cognitive maps to orient themselves in interaction, the same is true also of collectivities. The maps are part of their cultural store, accumulated over generations and, thus, heavily scented by the past" (1985: 101).

But at what cost do these rights come? This is a question the show poses through its unwillingness to bask in the moment of Jim's triumph. The episode chooses instead to contemplate the fate of the outsider in a society designed to exclude others. Anyone who might be supposed a threat is dealt with through violence. In short, the reality of the American Creed has come crashing down on

Hikmet Gersel in the most violent fashion. It is a spectacle that Det. Pembleton cannot let go without comment:

> Did you see what happened after the verdict was announced? They applauded. Those law-abiding citizens, those good-hearted people applauded the death of a child. Let me ask you a question, Tim, and then you tell me if it was racially motivated. If that kid had been American, if that kid had been white, do you think anyone would have cheered?

In one way, we can look upon the killing of Hikmet Gersel as the ultimate symbolic price a non-Westerner will pay for the failure to assimilate into Western culture. Hikmet's embrace of outdated cultural forms (Kiss) and his inability to communicate in the lingua franca of the New World Order seals his fate decisively and violently.

The final scene of "Colors" finds Jim not victorious, but isolated and removed, sitting in front of the television sadly contemplating a future in which his child might grow up to kill someone else's child: "No matter what I do for this baby, will he someday grow up and shoot somebody else's baby? Do me a favor, would you? Adjust the color. The green's a little off." Jim's statement hints at a certain hopefulness. In short, his perception of the world has changed; his reality is different now. His sadness stems from his realization that this change in perspective has come only at the cost of another human being's life. Returning again to Bolander's philosophical musings on reality, we are reminded that what colorblind people cannot see are differences in hue and, as Bolander suggests, perhaps this is the true reality. By suggesting that change is possible, the show offers up the suggestion that we might someday move beyond our loyalties to arbitrary, imaginary distinctions and toward the acknowledgement that we each have the same color blood.

What Jim Bayliss comes to represent is the danger lurking beneath America's conflicted image of itself as a community. His nice, suburban home with its manicured lawn, far removed from the violence and chaos of the inner city, bespeaks the legacy of "white flight" which left previously thriving urban spaces abandoned and wanting in the wake of the chimerical utopia of the American dream. Importantly, Jim confronts his fears about the future while sitting in front of his television. His request for a change of color is answered by the camera which, situated above and behind the appliance, cranes down and into the back of the set, blocking our view of the living room. We are left in a sea of black, hearing only the raucous cheering of a basketball game as the credits roll. It is a telling moment in which we are reminded that so much of what we know and imagine about this world and the others around us, what we take as given, what colors our perceptions and drives our actions, is inextricably bound up in the images we encounter every day from the safety of our living rooms.

Conclusion

Homicide represents a critical point in the cycle of police dramas between 1981 and 2000. It was easily one of the most socially and culturally critical series to play on commercial television. One of the means by which *Homicide* went about its critical work was to complicate the representational strategies of realism so central to the police genre. By using techniques and visual tropes borrowed from documentary realism (handheld cameras, location shooting, jump-cuts) in a highly self-conscious manner, the show called attention to the conventions and constructions that cling to representations of reality. In other words, *Homicide* asked us to acknowledge the process of signifying reality and, in so doing, undermined the generic conventions that are often used to instill a sense of authenticity and authority in documentary images – conventions that have informed the police genre from *Dragnet* to *COPS*. Once we acknowledge that the techniques of documentary realism are, like any other language system, arbitrary and based on conventions, we are in a position to question the authority of the text itself, especially as it relates to the maintenance of social order and moral authoritarianism so important to the genre.

In addition to its self-conscious commentaries on the conventions of realism, *Homicide* was also deeply concerned with the question of how communities are imagined, constructed, and maintained. Using Baltimore as the physical setting of the series gave this thematic concern specificity and weight. The questions that *Homicide* opened up about community were about more than geography; they were also about the *social construction* of communities – about communities as systems of social relationships. As questions about social relationships, they were posed in terms of race, class, and gender; and they were ultimately about interpretation and power.

In the end, *Homicide* was a highly self-reflexive but often contradictory text. Rather than simply telling a compelling tale of a crime and its solution, the program also considered its own place in the medium. It acknowledged the representational distortions that accrue in our living rooms and attempted to give a sense of how complicated issues of social and cultural conflict are packaged and presented for easy consumption by the media so that we might "make sense" of the world around us. At the same time, as David Simon pointed out, it actively participated in those distortions. One might even suggest that the series' focus on the impossibility of ever arriving at a simple truth positions it as nostalgic, like *T.J. Hooker*, for a past that worked. What *Homicide* illustrates as well as any other series is the range of narrative and representational possibilities open to any television text. In the world of narrative television, *Homicide* reminds us that there is nothing ideologically pure to be had – no simple sense to be made.

5

DO THE RIGHT THING

NYPD Blue and the Making of the Model Citizen

In the previous chapter's case study of *Homicide: Life on the Street*, I argued that police dramas engage in a symbolic construction of community. Because of the central relationship between community and the work of policing (to "Protect and Serve" the community), the establishment and maintenance of a sense of community is at the heart of television police dramas. Equally central is the related issue of citizenship.

Citizenship has always been one of the central thematic concerns of the American television police drama. From the no-nonsense procedurals of *Dragnet* to the stylized science of *CSI*, stories about the police are stories about citizenship, especially as it concerns individuals' relationships to the laws that govern the community (local, state, federal, international, etc.): the rights that those laws afford us and the responsibilities we have in light of those rights. Because the narrative structure of police dramas on television have tended to be episodic – focused on investigating and eventually solving a particular crime each week – the version of citizenship that they portray is what some critics have called "thin" or "shallow".[1] This "thin" version of citizenship is predicated mostly on the distinction between the *private* rights of choice-making individuals and their *public* responsibility "to general conformance to the rules of the community" (Bubeck 1995: 3). This position is typically associated with "liberal" or "neo-liberal" conceptions of citizenship in which the role of the state (in the form of laws, and the police and courts that uphold them) is seen as necessary, but only to the extent that it does not interfere with *private* individual freedom: the freedom to live as one wishes within the general structure of the community's laws (Bubeck 1995: 4; Faulks 2000: 11). The majority of episodic police dramas fall into this category of "thin" citizenship through their emphasis of crime as an individualized breach of the community's rules, and punishment as a way of protecting the rights of innocent individuals.

With the shift toward seriality and melodrama in the 1980s and 1990s, a number of the police dramas that I have discussed previously (e.g. *Hill Street Blues* and *Homicide*) seemed to shift away from a predominantly neo-liberal version of citizenship in favor of an expanded view that focused largely on the private, intimate, and often intense interpersonal relationships among and between police officers, victims, and suspects. These series connected the public performance of citizenship to private interpersonal interactions by highlighting the role of relationships as an important indicator of one's moral status within the community. This chapter examines one such series, *NYPD Blue*, as an example of the way in which serial police dramas have used the heightened moral concerns of melodrama in order to explore questions about what constitutes "good" or productive citizenship. Specifically, I explore how the melodramatic mode opens up and expands questions about rights, responsibilities, and the consequences of our decisions: each central to discourses about citizenship. I begin by placing the series in its institutional context, as it courted controversy by offering more "realistic" and "adult" stories (in terms of language and sexuality) that delved more deeply into the intimate lives of the police. I then turn to explore the relationship between realism and melodrama in the series – the way in which the series uses contemporary social discourses to explore the concept of virtue so central to melodrama. Finally, I explore the ways in which the series uses this melodramatic mode to construct an image of "intimate" citizenship that nominates the figure of the detective as the model citizen: responsible for oneself and *obligated to others*. This focus on a more "intimate citizenship" illustrates the ways in which the increased focus on the private lives of the police has not stunted, but rather *extended* the genre's concern with citizenship as its primary thematic focus.

Conflict Makes Drama

When *NYPD Blue* premiered on September 21, 1993, it had already made more headlines than most established series. Almost all of the talk about it focused on producer Steven Bochco's promise to deliver television's first R-rated series.[2] According to an ABC press release, the series promised "a unique blend of realism and smoldering situations," which translated mostly as increased violence, rough language, and heightened sexuality, including partial nudity during scenes of sexual intimacy. This all promised to be too much for Donald Wildmon, founder and head of the American Family Association. Without the benefit of seeing the series' first episode, Wildmon took out full-page ads in *USA Today* calling for concerned citizens to join his cause by signing a petition addressed to advertisers, affiliates, and the ABC network ("Wildmon Targets Cop Shows": 42). By the time of the premiere, Wildmon had convinced fifty-seven ABC affiliates to refuse to air the pilot episode and forty-four of these affiliates to continue their boycott of the series into the remainder of the season ("ABC Stations Feel 'NYPD' Heat": 78).[3] Because of this threat to the number of potential viewers, and the added threat of boycotts against their own products, many prominent advertisers steered clear of the series as well (Leland 1993: 56; Mandese 1993: 1).[4]

This initial note of controversy set the tone for the way many viewers and reviewers approached the series. Interestingly, a number of viewers and critics felt that the controversy had been overblown once they actually saw the first episode (Mandese 1993: 8). Richard Zoglin betrayed a whiff of disappointment when he stated in an early review: "It is a sign of how placid the rest of network television has become that Bochco's strong but relatively conventional cop show has incited such an outcry" (1993: 81). Similarly, James Wolcott criticized Bochco and the series for sticking too close to the conventions of the 1970s-era anti-authoritarian cops and updating that cycle only through a stylish addition of "grit glued to its surface" (1993: 218). Frank McConnell, writing in the Catholic publication, *Commonweal*, argued that, while *NYPD Blue* was exceptionally good television, the controversy surrounding it had imbued it with an air of importance that it may not have truly deserved:

> Steven Bochco must be, at this moment, one of the happiest guys in the world: not only because he's brought forth yet one more splendid demonstration of how good TV can be, but because he's summoned forth the ideal, thin-lipped and school-marmish adversary to make his grand show look even better than it really is.
>
> *(1993: 21)*

Reactions such as these hinted that controversy is precisely what Steven Bochco and David Milch wanted for the series – a clever way to meet the challenges of marketing a series in an era of expanded viewing options.[5]

Bochco, of course, has always courted controversy openly, as he demonstrated repeatedly during his famed fights with NBC's Standards and Practices department while producing *Hill Street Blues*.[6] But the decision to push the boundaries of taste and subject matter on *NYPD Blue* likely had more to do with the realities of the television market than with his desire to torment network executives. As he told James Longworth in a recent interview: "I felt if we didn't make this kind of show, and get people back in our tent by being more adult, and more contemporary in our use of language, then we were going to be out of business; it's as simple as that" (Longworth 2000: 202). The phrase "out of business" refers to both the traditional networks who faced increasing threats from cable and satellite, as well as Bochco's production company, which needed to create a series that would survive long enough to move into syndication. His comments hint at his prescient understanding of the changing market for syndicated programming, which now includes prominent cable networks such as CourtTV, TNT, and fX, that are not limited by the same FCC content restrictions as broadcasters and can thus capitalize on risky (and risqué) programming.[7] Indeed, each of these cable networks has shown *NYPD Blue* in syndication.

The controversy over *NYPD Blue* died rather quickly as a result of its resounding success in the ratings. The first episode received a national rating of 15.4 and a 27 share, which placed the series in the top ten (Elliott 1993: D19). By the middle of the season, the series was commanding an audience share in the mid-twenties (Milch 1995: 91). *NYPD Blue* finished seventeenth overall in the ratings during its first season, second in hour dramas behind *Northern Exposure* (Brooks and Marsh 1999: 1257). The series also received twenty-six Emmy nominations, more than any other series in history. Though Wildmon continued his boycott efforts into the second season, advertisers and affiliates alike decided to accept the series, and the terms for understanding its content began to shift: from scandal to "quality."[8] Additionally, as series like *The Sopranos*, *Sex and the City*, *Six Feet Under*, and *Queer as Folk* began to experience wild success on the premium cable networks HBO and Showtime, the raciness of *NYPD Blue* paled in comparison. Since these pay-cable networks are free from FCC indecency rules, these series could push boundaries of language, violence, and sexuality that *NYPD Blue* could only dream of approaching.

Given its surprising success, one might argue that *NYPD Blue* proved that the hour-long drama, considered a relic, could once again be a profitable ratings success.[9] Shows like *ER*, *Judging Amy*, *Providence*, *Chicago Hope*, and *The Practice* (to name only a few) followed *NYPD Blue* to ratings success and represented a resurgence of the hour-long drama as a central part of network programming. This resurgence also served as a central component of cable networks' forays into original programming and their strategy for buying off-network rights for series. As an example of this trend, TNT redesigned its image as "The Place for Drama" by investing in the off-network rights for dramas such as those mentioned above. In order to publicize its turn to drama, the cable network began airing promotional spots that included actors describing what constitutes drama: "Conflict makes drama." TNT even created an entire promotional segment specifically for *NYPD Blue* that constructed a more specific thematic discourse for the series: "*NYPD Blue* is about the redemption of Sipowicz."

These discourses of conflict and redemption are indeed borne out by the arc of Detective Andy Sipowicz's trials over the course of twelve seasons. In the pilot episode alone, he lies on the witness stand in court, verbally abuses the Assistant D.A., drinks himself into a stupor, physically assaults a government witness, and gets shot. Over the course of several seasons, Sipowicz experiences health problems, struggles with alcoholism, and copes with the deaths of his eldest son, a wife, and two partners. Given the early depiction of Sipowicz as a dangerous loose canon – violent and temperamental, racist, sexist, and alcoholic – the moral and emotional education of this character seems a natural fit for the focus of the drama.

Actually, the emotional lives of almost *all* the characters are on display far more often than the controversial "blue" language and partial nudity. These are characters in crisis – personal and physical – and they deal with these crises through emotional displays of suffering and sacrifice. At their core, the characters of *NYPD Blue* are, as David Milch points out, "self divided." The series as a whole "is intensely about the question of how to be and that's what governs all the stories" (Longworth 2000: 93). If, as the TNT promo states, conflict makes drama, then this question of "how to be" defines the central conflict of *NYPD Blue*.

The question of "how to be" is, of course, a question about citizenship in the broadest sense: of how to balance one's rights and responsibilities in the decisions one makes and the actions one takes. It is also a moral question and places the series squarely in the arena of melodrama, which presents what Peter Brooks calls "the social world as the scene of a dramatic choice between heightened moral alternatives" in which these moral alternatives are constructed as "large and basic ethical conflicts" (1976: 5–6). At times, the characters of *NYPD Blue*, like almost all television detectives, can seem almost preternaturally virtuous in their nearly unerring sense of right and wrong. But the means by which they pursue justice are often morally and ethically compromised. As one reviewer described the character of Sipowicz: "Balding, overweight, a recovering alcoholic and unreconstructed racist, he is a complicated moral center for a network TV series – an obsolescent white guy with a powerful sense of right and wrong but a flawed compass" (Leland & Flemming 1993: 57). This metaphor of the flawed compass underscores the point that *NYPD Blue* is not so much about the weekly reaffirmation of a singular legal definition of social justice and proper citizenship, but about the constant, sometimes aimless, search for any sense of justice and personal direction in a complicated moral landscape.

The work of policing provides a useful site for addressing these issues for two reasons. First, it provides a realistic setting of emotional and physical peril for the characters. Second, it provides access to a range of criminals, suspects, and victims, against whom the detective-heroes may be measured. In the end, the episodic confrontation with this range of criminals and suspects serves the larger narrative arc of the detectives' personal growth: just as the detectives work to contain social chaos by closing cases, the stories of *NYPD Blue* work to discipline its characters' emotional lives – to teach them control in the face of personal chaos but also to teach them the importance of knowing when to relinquish that control by reaching out for help. Through this dialectic of introspection and "reaching out," the series forges an image of the model citizen: disciplined and self-reliant, but also open and empathic. By focusing on this idea of a *model* citizen, my argument emphasizes that these characters are not static *ideals*, but rather flawed characters who are always in the process of trying to understand and cope with their weaknesses and helping others in this regard as well.

Melodrama: Realism, Virtue, and Moral Action

Three key aspects of melodrama as a form of narrative deserve particular attention, and help to shed light on the ways that *NYPD Blue* (and police dramas in general) constructs a moral world in which a vision of model citizenship takes shape.

The first is the relationship between melodrama and realism that has been a point of ongoing disagreement and misrecognition among critics. The second issue is the centrality in melodrama of a return to virtue and innocence through moral action. Finally, we need to consider how the series constructs moral action itself as a "victory over repression" (Brooks 1976: 4). These threads are not mutually exclusive; they cross over and tangle with one another in many ways. Still, there is heuristic value in separating them for the time being as a way of charting the melodramatic field of *NYPD Blue* that supports its particular construction of citizenship.

As I have discussed in previous chapters, one of the key discourses within the police drama is its insistence on realism – both in form and subject matter. *NYPD Blue* has embraced this realist aesthetic wholeheartedly, from the now nearly obligatory handheld camerawork, to its necessarily contemporary subject matter, to the portrayal of the emotional lives of the cops themselves. At the same time, the use of primarily interior settings points to the fact that the realism in the series is mostly a product of subject matter and character than of *mise-en-scène* and physical action. This more sociological realism has been lauded by critics as a key to the series' success. Elayne Rapping has noted that the series "gets the nuances of contemporary social and psychological interactions just right" (1994: 37) and Frank McConnell enthused that "the big-city ambiance perfectly captures the massive, sudden loneliness, and metaphysical finality of *Urbs Americana* [sic]" (1993: 20). David Thomson, writing in *Esquire,* labeled *NYPD Blue* "a tough, gritty, naturalistic series" and argued that you "can feel the scum on the streets the way you can feel matters of race, poverty, and prejudice" (1998: 65). As for the acting, Thomson referred to Sipowicz (played by Dennis Franz) as "one of the few people you can smell on TV" (1998: 66), and David Milch reported that Bill Clark, the ex-cop who serves as one of the producers of *NYPD Blue,* considered David Caruso to be "100 percent believable as a cop" (1995: 47).[10]

While the series was an immediate success with critics on the whole, the melodramatic elements of the series have not always been appreciated. One early critic summed the show up this way: "The show, to judge from the first episode, is a generic cop melodrama about divorce and alcohol and middle-age burnout" (Bowman 1993: 54). Similarly, a critic at *The Boston Globe* labeled the series "overblown character melodrama" (Gilbert 1997: E1); a critic at *The Boston Herald* indicated that the "high melodrama" and "theatrics" somehow managed not to overwhelm otherwise "graceful, gutsy storytelling" (Collins 1998: O51); and a critic at *The Atlanta Journal and Constitution* referred to the monotony of the series' "Nowheresville station-house soap-operas" (Jubera 2000: 1C).

These misgivings about the emotionalism and melodrama of *NYPD Blue* are in line with larger criticisms of melodrama as a cultural form, especially as they articulate gendered distinctions between classical realism and the theatrics of "high melodrama." As Peter Brooks (1976), Christine Gledhill (1992), and Linda Williams (1998) have pointed out, the very idea of melodrama has been criticized as essentially "anti-realist" and excessive in its emotionalism – qualities that relegated it to the lower depths of "feminized" mass entertainment. Only when the excessive emotionalism was seen as ironic, as in the much-lauded films of Douglas Sirk, was melodrama "'redeemed' as a genre" and these particular criticisms reversed (Williams 1998: 43). Writing specifically about film studies and melodrama, Williams notes that critics, in their embrace of Sirk, "established a rigid polarity: on the one hand, a bourgeois, classical realist, acritical 'norm,' and on the other hand, an anti-realist, melodramatic, critical 'excess'" (1998: 44). Even in this more positive formulation, however, melodrama is still understood as something *other* than the realist (masculine) norm.

What Williams argues, however, is that melodrama has been misidentified as a category of the "woman's film" or "woman's program" (such as soap operas), which are often treated as inadequate attempts at realism: too overblown and emotional, too indebted to "simplistic Victorian personifications of Good and Evil, Innocence and Villainy" (Gledhill 1992: 32). For Williams, Gledhill, and Brooks, melodrama is more fruitfully and accurately understood as a central mode of American storytelling as opposed to a mere sub-category:

> If we want to confront the centrality of melodrama to American moving-image culture, we must first turn to the most basic forms of melodrama, and not only to a sub-ghetto of women's films, to seek out the dominant features of an American melodramatic mode. For if melodrama was misclassified as a sentimental genre for women, it is partly because other melodramatic genres such as the western and gangster films, which received early legitimacy in film study, had already been constructed...in relation to supposedly masculine cultural values.
>
> *(Williams 1998: 50)*

To understand melodrama as a central mode of storytelling rather than a debased and anachronistic genre is to engage with what Brooks calls the "melodramatic possibilities" of a range of genres, including the police drama which is typically understood as primarily masculine and realist (Brooks 1976: 204). By engaging with the melodramatic possibilities of the police drama we can open up our understanding of it as something other than obsessively conservative in its drive toward justice and narrative closure. Instead, focusing on melodrama opens up opportunities to discuss the "intense inner drama" of the series "where every gesture, however frivolous or insignificant it may seem, is charged with the conflict between light and darkness, salvation and damnation" (Brooks 1976: 5).

The critical division between realism and melodrama is misleading in that the distinction is based primarily on formal qualities such as acting, psychological motivation, and *mise-en-scène*: the surface indicators of verisimilitude. As Linda Williams argues, however, realistic *effects* serve melodramatic *affects* (1998: 42). Melodrama, rather than simply being a theatrical style of presentation, is a mode of representation concerned with moral divisions, and it must engage contemporary issues in a realistic manner if it is to have any resonance with audiences – that is, if it is to achieve *affect*. Indeed, as Gledhill insists, "modern melodrama draws on *contemporary discourses* for the apportioning of responsibility, guilt, and innocence – psychoanalysis, marriage guidance, medical ethics, politics, even feminism" (1992: 32). More specifically, Peter Brooks argues that, in addition to its focus on formal realism, the police series historically has offered one of the "clearest possible repertories of melodramatic conflict" (1976: 204). Traditionally, this conflict has been staged between clearly drawn "villains and heroes (who can often be recognized simply by uniform), of menace and salvation" (Brooks 1976: 204). But the impulse toward even greater psychological realism through more complex portrayals of the police officer him/herself does not negate the melodramatic thrust of the series; it simply extends and modernizes it.

> That [police series] have become increasingly "psychologized" – that cops must be experts in human relations and bad men are quasi-Dosteoevskyan figures – in no sense violates the melodramatic context. It is not that melodramatic conflict has been interiorized and refined to the vanishing point, but on the contrary that psychology has been externalized, made accessible and immediate through a full realization of its melodramatic possibilities.
>
> *(Brooks 1976: 204)*

Rather than ridding the police series of its impulses toward melodrama, the push toward greater psychological realism merely extends those possibilities in the genre, opening up new, more complex and contemporary areas for conflict.

As an example of how contemporary discourses drive melodramatic conflicts in *NYPD Blue*, we can consider the centrality of racism in the series. David Milch has said: "I wanted *NYPD Blue* to be about race. One of its most important aspects was the opportunity to address race, among other issues, in the kind of complexity I hadn't seen in other television drama" (Schiff 1997 5). The character of Andy Sipowicz provides a useful illustration of the kind of complexity that Milch is talking about. While Sipowicz is prone to voicing racial prejudices, the series avoids simple admonitions of his character. Instead *NYPD Blue* uses Sipowicz's racial biases as a way to grapple with the contradictions that frame discourses about race in urban America: where the responsibility for racism lies, and what might be done about it.

In a pair of episodes from the third and fourth seasons ("The Backboard Jungle" and "Where's 'Swaldo?" respectively), Sipowicz is forced to confront his racial prejudices and does so with only partial success. In the third-season episode, Sipowicz and his partner, Bobby Simone, try to get a man named Kwasi, a prominent black leader in the community and a witness in a murder investigation, to come with them to the station:

KWASI: (to Sipowicz) I don't have to go anywhere with you. You're dealing with the one nigger in a thousand who knows what you can and cannot do.

SIPOWICZ: I'm dealing with the *nigger* whose big mouth is responsible for this mess! (Sipowicz punches the epithet in his delivery).

This response to Kwasi is overheard by a reporter and, of course, destroys any opportunity that Sipowicz and Simone may have had for getting help with their investigation from the black community. It also ignites tempers and raises tensions between Sipowicz and other members of the squad. Simone is especially troubled by Sipowicz's unwillingness to address his prejudices: "Partner, I am not comfortable with those words. I am not comfortable with the thoughts behind them. You gotta understand that." This final statement could be used as the object lesson for the episode: the lesson that we are supposed to learn from the more sensitive and enlightened figure of Simone. But the scene resonates far beyond that simple lesson.

The conflict between Sipowicz and Simone illustrates what Brooks calls the "drama of moral consciousness" (1976: 6) in which the drama exists not in outward actions but inward apprehension. The conflict is made manifest in the effect of Sipowicz's racism on his relationships (with victims, co-workers, and his wife), but it also registers *within* Sipowicz as well. His real struggle is internal and is played out in the second episode of the arc. In this episode ("Where's 'Swaldo?"), Kwasi has been found murdered and his wife points out Sipowicz to her young daughter as the man who called her father a "nigger." The final scene begins with Sipowicz sitting alone in the squad locker room when Simone enters. "Did you hear her tell that little girl to hate me?" asks Sipowicz. He then embarks on a story about returning from Vietnam and joining the police force, working undercover in order to infiltrate the Black Panthers. He ends the story with an angry tirade in which he refers to the Panthers as "spades." Simone berates Sipowicz for using that term and Sipowicz responds with a long soliloquy about how he got to this emotional place:

> It was hard for me. You understand that? Come back to the world...at least I thought I could talk to someone, and maybe somebody respects what I did. I dreamt of being a cop. And now when a cop sees me on the street he spits on the ground. And I know he's supposed to because he sees me kissing bastard's asses want to blow up the bank that my mother pays our house loans at. And I'm telling these bastards how brave and great I think they are – how I loved it when my dad, he finally saves enough after *he's* in the service to move us from the Quonsets and we're finally in a decent neighborhood until *they* move in. And I have to fight to keep my lunch money and the project turns into a sty. And he gets his eye put out by one of them, drunk with a hammer, who don't want his gas meter read. To have her tell that little girl to hate me...I try to do my job, put your feelings aside unless they show you they're wrong. And that little girl is told to hate me.

Simone responds to Sipowicz's tirade by confronting the self-pitying note that his partner has struck:

SIMONE: What is she supposed to do, Andy? She supposed to feel sorry for you? She supposed to think it's OK you callin' her father a nigger because you had bad times growing up?
SIPOWICZ: I've had bad times from those people my whole life.
SIMONE: And my dad got his ass kicked at shake-ups by white longshoremen for being dark-skinned. I don't hate them. I don't hate you.
SIPOWICZ: I'm supposed to care what happened to her people three hundred years ago?
SIMONE: There you go....
SIPOWICZ: At least she shook my hand. I don't think *she* hated me that much.

Simone leaves the locker room and the episode ends as the camera lingers on Sipowicz, sitting on a bench, staring at the wall. The scene is grounded in recognizable experiences and the social realities such as the hostility that Vietnam veterans experienced upon returning home and the swirling racial tensions that defined urban life in the late 1960s. But it is melodramatic in the sense that it is about "inward apprehensions" – the moral consciousness that is "played out under the surface of things" (Brooks 1976: 4). Surface gestures, such as a handshake, ignite a conflict within Sipowicz and ask the audience to also engage in the same kind of conflict and introspection that Sipowicz must now undergo. Milch underscores his point when he states: "I've been called a racist, and I think it's easier to call me a racist, or to call the show racist, than to call Sipowicz a racist because people like Sipowicz" (Schiff 1997: 5).

In defending his portrayal of Sipowicz's racism on more basic terms, Milch explains it's also simply a matter of realism: "*not* to portray racists in a series about New York cops would have made the show incredible" (1995: 184). But the emphasis on verisimilitude alone ignores the melodramatic conflict of the scene, or what Brooks calls the "moral occult...the domain of operative spiritual values which is both indicated within and masked by the surface reality" (1976: 5). The rub here, of course, is that Sipowicz is the central heroic figure with whom we are asked to identify. Regardless of the scene's verisimilitude, how are viewers supposed to reconcile Sipowicz's inability to come to terms with his racist language and his position as a virtuous hero? The answer to this question lies within the melodramatic mode itself. As Linda Williams points out, "the basic vernacular" of American narrative culture "consists of a story that generates sympathy for a hero who is also a victim" (1998: 58). Of course, we are not asked to see Sipowicz as a victim of social change – the circumstances that altered the fabric of his neighborhood and fueled the Black Panthers. Instead, we are asked to see Sipowicz as a victim of his own inability to come to terms with his hatred and racism. Sipowicz is a victim of his own consciousness which must "purge itself" in order to attain virtue (Brooks 1976: 5). Our ability to reconcile his racism with his status as hero, therefore, rests on our ability to recognize his status as a victim first, who will work hard to liberate his/her virtue from the forces that "repressed, expulsed, silenced it, to assert its wholeness and vindicate its right to existence" (Brooks 1976: 42).

In the melodramatic mode, this victimization is typically purged either through "a paroxysm of pathos" or through physical action, or often a combination of both (Williams 1998: 58). Sipowicz's monologue is pure pathetic expression – an outpouring of feeling, however troubling – versus Simone's clipped and disciplined responses. Peter Brooks argues that the expressive language of melodrama "acts as a carrier or conduit for the return of something repressed, articulating those very terms that cannot be used in normal, repressed psychic circumstances" (1976: 42). Sipowicz's tirade is a way of "getting it out," not just for the character but for the audience as well. Furthermore, because melodramatic conflicts such as this turn "less on the triumph of virtue than on making the world morally legible" (Brooks 1976: 42), the scene is not designed as an opportunity for Simone to teach Sipowicz a moral lesson about racism once and for all. Instead, it is designed to chart the terrain of contemporary racism that "political dogma cannot articulate" (Gledhill 1992: 33). Once Sipowicz has expressed his feelings and articulated his status as the victim of these feelings, he can begin the process of dealing with them in a more productive manner. When Sipowicz utters his final words in the scene ("I don't think *she* hated me that much"), he has, in effect, located the possibility of virtue in the form of the young daughter. He has begun the process of no longer naming himself as the victim of race but, instead, seeing the young girl as the potential victim whose innocence and virtue (demonstrated in her willingness to shake Sipowicz's hand) stands the best chance of overcoming the destructive racism that Sipowicz represents.

The scene also illustrates another important element of the melodramatic mode: "the wish-fulfilling impulse towards the achievement of justice…as the powerless yet virtuous seek to return to the 'innocence' of their origins" (Williams 1998: 48). This innocence is, of course, located in the daughter who cannot comprehend the depth of Sipowicz's racism or even locate its origins. By trying to identify with her, Sipowicz works toward relocating that moment of innocence in himself before he, too, became a victim.

This desire to return to a state of lost innocence is manifested in a significant number of episodes of *NYPD Blue* that present children as victims who are either abused or murdered. These episodes provide some of the most direct opportunities for melodrama in the series: the recognition of virtue played out through the suffering of the innocent. For example, in an episode in the second season ("Simone Says") a woman, Mrs. Davis, shows up in the squadroom looking for a detective to talk to her husband, whom she suspects is sexually abusing their 14-year-old daughter. The husband admits to abusing his daughter and a scene in which the husband and wife accidentally confront one another in the squadroom erupts into the kind of gnashing of teeth commonly associated with high melodrama: what Linda Williams calls the "sensation scene" (1998: 59) that is generated by the realistic cause and effect narrative but provides an expressive outlet for the "grandiose moral terms of the drama" (Brooks 1976: 8). The mother's violent outburst is understandable at the level of the plot alone. But it also reverberates outward and links to a deeper moral discourse about the need to protect the innocent.

These moral terms of the drama are underscored in the penultimate scene of the episode in which Simone and Sipowicz have a drink with fellow detectives, Medavoy and Martinez, and bond over their respective hobbies. During the course of their conversation, Simone reveals that he raises pigeons and talks about how reliable they are in their instinct to return to the roost. Sipowicz responds with a story about his fish:

> I got a clownfish couple that just had eggs. In the morning when I'm having my coffee, the male, he cleans each egg with his mouth. Never breaks one. And then the whole day while I'm working, him and the wife are guarding that nest. Fanning water all over the eggs. Those are dedicated fish. You see that kind of thing in pigeons?

"Pigeons make good parents," replies Simone. The scene accomplishes two things: first, it serves the development of character relationships so important to serial drama by providing Sipowicz and Simone with a point of connection; second, it returns us to an image of redemptive virtue so central to the melodramatic mode. The story works not only to establish a bond between Sipowicz and his new partner, but also to respond to the story of neglect and abuse that has just played out in the station. The suffering of the innocent daughter and the virtuous action of the fish are separated (they are not found in the same being) but they are linked to one another metaphorically. Sipowicz, as the vehicle of the metaphor, locates the potential victim/hero in the figures of the detectives who must bear witness to tragedy and take action. The suffering victim is not only the child, but the detective who must try and find a way back to a state of virtuous innocence, if only metaphorically, in the face of constant evil.

No episode better illustrates this desire for the return of lost innocence than the two part "Lost Israel" (season five). The primary story in this episode centers on a missing 8-year-old boy, named Brian Eagan, and his parents who are worried that he may have been raped and killed by a homeless man named Israel. The detectives bring Israel in for questioning only to find that he cannot speak. He understands their questions but cannot respond with his voice. Instead, he points to passages in a copy of the Old Testament he carries with him. Neither Sipowicz nor Simone can make sense of these messages and dismiss them as nonsense. Israel is childlike and frightened of the detectives who are quickly convinced of his innocence in the matter but decide to keep him in a cell as a way to mislead Brian's father, whom they have become convinced is the guilty party. Mid-way through the episode, Israel is found dead in his cell, having committed suicide. He has left his bible open to a passage that Sipowicz still cannot understand.

Relatively early in the episode, the boy is found strangled and it is confirmed that he has been sexually abused over a long period of time. Sipowicz and Simone spend the remainder of the episode trying to prove the father's guilt. One way they do this is by having Det. Diane Russell forge a friendly and trusting relationship with the wife in an effort to get her to implicate her husband. As the circle closes in on the father, he finally confesses his guilt to Simone in a convoluted monologue:

> I am not so presumptuous or egotistical to pretend that I communicate with God. What I do know is what's necessary for life not to become a grotesque insanity and to protect a child that you love. I know the burden of what has to be done. [Background music begins] Because previously you can't forgive yourself. And you'd beaten that in him. You can't forgive your own erection lying next to your counselor and you know you must be foul. And so you beat yourself and your boy. I don't presume to communicate with God. But can you believe in any God that didn't want his child protected when he sees him crying at night and wetting himself. And some monster making him lie, hide and be so sad and afraid. How wouldn't God want him taken away!? [He makes his hands as if to strangle and then looks in horror at the recognition of his own capacity for evil].

Like the accidental meeting between the mother and father in the earlier episode, this scene has all the marks of a traditional melodrama, with its ominous music, overwrought emotional tone, and its display of abject horror on the faces of the participants. But it is also an intensely moving scene; it asks the viewer to at least try and comprehend the origins of the father's monstrosity, as pathetic as they might seem. The scene "draws into a public arena, desires, fears, and identities which lie beneath the surface of the publicly acknowledged world" (Gledhill 1992: 33). It dredges up elements of ourselves that we would prefer not to acknowledge so that we might chart more clearly the moral terrain of our society.

The confession scene also illustrates the third element of melodrama: the insistence on a victim who is capable of moral action by overcoming repression. The father is himself a victim of abuse but he is incapable of taking moral action. He has simply repeated the monstrous and long-repressed actions of his own father. The burden of moral action in the episode is left to his wife. At the conclusion of the episode, after the father has confessed, Simone finds himself at odds, pacing angrily around his desk, unable to corral his emotional response to the father's confession:

SIMONE: There is no forgiveness for what he did. He is gonna burn in Hell for this! But first I want him to live a good long time so he can wake up crying and screaming all night realizing what he did. He wants me to shoot him? I hope that was him knowing that there is no forgiveness for that.

SIPOWICZ: He asked you to shoot him? He was asking the wrong guy.

Det. Russell and the mother enter the squad room, interrupting the tirade. The news that her husband has confessed to killing Brian prompts Mrs. Eagan to decide that she, too, must confess to her own culpability in Brain's death:

MRS. EAGAN: I must have collaborated. I have to confess it. I must not have wanted to protect Brian.

SIPOWICZ: Sometimes, people that do something horrible know how bad it is and they hide it very good.

MRS. EAGAN: Thinking we'll die if we believe it doesn't excuse not believing. And maybe we *should* die.

RUSSELL: Brian would have no one to remember him then. You have to keep remembering and loving him: remembering all his goodness; remembering all the joy he brought you.

MRS. EAGAN: I don't deserve to remember him. I failed. He needed me to protect him and I couldn't. How can I live with myself after that?

RUSSELL: You have to.

Sipowicz then picks up Israel's Bible and reads the passage that Israel had marked before his own death. The passage is from Psalm 119: "When wilt thou comfort me? For I am become like a bottle in the smoke. Do I not forget thy statutes? How many are the days of thy servant? When wilt thou execute judgment on them that persecute me?" The passage emphasizes the need to hold fast to God's precepts in the face of persecution; it insists that human existence is fraught with trials of the spirit and that these trials will be painful – a trial that we do not feel is no trial at all. Importantly, the passage assures us that these trials will not consume us. Like a "bottle in the smoke," we are surrounded by trials but not consumed by the fire that creates the smoke.[11]

The action that Mrs. Eagan must take is to endure the trial of her son's death by simply living, recovering, and continuing to remember Brian. The biblical passage itself neatly encapsulates the moral basis of melodrama: that through the virtuous action of suffering and sacrifice and the expression of that which is so deeply repressed in us, we can recover or protect the state of innocence to which we all desire a return. Through the expression of deeply held grief, Mrs. Egan is able to take action; and it is through the display of this action that the episode achieves the affect so central to the melodramatic mode. This affect doesn't simply underscore clear distinctions between good and evil. Rather, it highlights the dialectic relationship between the two and emphasizes the ongoing effort required to achieve the virtue that is the primary concern of melodrama.

Model Citizenship: Obligations and Intimacy

Having illustrated the ways in which *NYPD Blue* fits within the melodramatic mode, we can now turn to the issue of how the series addresses its central thematic concern: "how to be." This is a question of both individual morals and broader social responsibilities. Put another way, it's a question of citizenship: of how individuals should balance their rights and responsibilities as members of a community. In this section, I will illustrate how *NYPD Blue* engages in a complex dialog with other popular cultural representations of law and order, constructing a model of "good" citizenship that is located squarely in the realm of interpersonal relationships. *NYPD Blue* constructs a model citizen who is self-reliant and self-aware, but who is also a member of a community capable of "reaching out" in empathy and solidarity with others who share a sense of civic obligation. The heroes of *NYPD Blue* work for the good of their community not only through their roles as police detectives, but through their network of close and productive interpersonal relationships. Thus, *NYPD Blue* constructs the model citizen not through privatized action, but through productive social *inter*action.

The previous example of Mrs. Eagan's confession illustrates what productive social interaction looks like on *NYPD Blue*. Mrs. Eagan's ability to take moral action is predicated on the existence of support from those closest to her, represented here by the emotional and physical support she gets from Simone, Sipowicz, and Russell. In this scene, the principal characters exhibit their sense of obligation to others. The detectives move beyond their duties as police officers in order to assume their social obligations to Mrs. Eagan as an individual. For her own part, Mrs. Eagan's willingness to confess underscores her sense of obligation to her son. The act of confession is central to *NYPD Blue* and is a key to understanding the way in which the series constructs its model of citizenship.

From Legal to Moral Responsibility

The issue of citizenship is almost always central to those narratives that deal with the search for justice, such as police dramas, courtroom dramas, and the rash of courtroom reality shows like *The People's Court*, *Texas Justice*, and *Judge Judy*. In programs about justice, the issue of personal responsibility is frequently nominated as the key to proper citizenship. Culpability and guilt are often located at the level of the individual. Justice is fulfilled when the guilty or culpable individual is punished, and redemption is located in the individual's ability to accept the consequences of his or her choices and actions. The conflict that drives these dramas often exists precisely in the *unwillingness* of individuals to consider or recognize their responsibility, or in their efforts to *hide* their culpability (usually unsuccessfully). The role of the Law (in the figure of the police officer, detective, lawyer, and judge) is to re-inscribe the boundaries of good citizenship by exposing the individual responsible for crime and making that individual abide by the consequences of his or her actions in real, material ways through fines, prison, or even death.

The narratives of programs like *Judge Judy* offer what Laurie Ouellette has called the "neoliberal technology of everyday citizenship" (2004: 232). Neo-liberalism is most closely associated with economic policies that work to replace the welfare state in favor of the rule of the free market (Martinez and Garcia 1997: 1), and the ideas of personal choice and responsibility are an extension of this market philosophy. Programs like *Judge Judy* extend this market philosophy to the realm of justice and the very notion of citizenship itself, marking "good" citizenship as an extreme individualism that emphasizes, above all, individual rights. Because of neo-liberalism's focus on individual rights, it posits a model of citizenship in which one is primarily responsible for and to oneself. A corollary to this liberal view of citizenship is the more socially conservative concept that places emphasis on responsibility to the community. According to this view, neo-liberalism's focus on individual rights leads not to liberty but to license (Faulks 2000: 70). Social conservatives, in their efforts to rejuvenate citizenship as a central component in the maintenance of communities, "reconnect rights to responsibilities by making entitlements dependent on the performance of duties" (Faulks 2000: 70). But, as Keith Faulks argues, this notion of citizenship as responsibility often treats the community as an abstract concept, neglecting the structural and material contexts of everyday life that render vulnerable many whose social and political rights have yet to be fully realized such as women, minorities, homosexuals, and children.

As a corrective to this division between rights and responsibilities, Faulks argues for a more "holistic" notion of citizenship that creates a balance between the two. In particular, Faulks complicates the issue of responsibility, discriminating between two types: duties and obligations. "Duties may be seen as those responsibilities imposed by law and carry some kind of sanction if the individual does not honor them. Obligations, in contrast, may be seen as voluntary and as an expression of solidarity and empathy with others" (2000: 82). For Faulks, this focus on obligation results in a more "intimate citizenship" which applies "the principles of citizenship to interpersonal relationships" (2000: 124). According to Anthony Giddens, "individuals who have a good understanding of their own emotional makeup, and who are able to communicate effectively with others on a personal basis, are likely to be well-prepared for the wider tasks of citizenship" (Faulks 2000: 126). In speaking of the "wider tasks of citizenship," though, Giddens remains committed to the notion of politics as a public ritual: a function of the state (Faulks 2000: 127). Faulks's call for a more holistic citizenship is meant to collapse the rigid boundaries between public and private life that have defined traditional notions of citizenship.

The call for a model of citizenship that is based at least in part on intimacy is a response to the neo-liberal concept of citizenship described above. As Ken Plummer argues, traditional notions of citizenship "cannot readily accommodate the very categories of difference and consequent inequalities and patterns of exclusion that have been the focus of so much recent thought" (2003: 52). More specifically, feminist critics have taken explicit aim at the public/private divide that has structured debates about citizenship. Ruth Lister, for example, argues for a critical "rearticulation" of the divide: "This rearticulation involves the de-gendering of the public-private divide; acknowledgment of the ways its two sides interact with each other; and recognition of its fluid political nature" (2003: 143). In particular, critics such as Lister advocate strongly for the inclusion of domestic work and *care* as central to any concept of citizenship since "the public exercise of citizenship has always depended" on them (2003: 120). On the point of *care*, Diemut Bubeck points out the ways in which "private virtues and skills arising from and informing the practices of care can be argued to be relevant and important in citizens, too" (1995: 26). For Bubeck, a "thicker," more active and useful form of citizenship emerges when we transcend or shift the boundaries between public and private. As Faulks points out, this shift also alters the terrain of citizenship, changing it from a purely legal identity to a *moral* identity. It is precisely this more intimate form of citizenship that *NYPD Blue* explores and eventually constructs through its portrayal of both the professional and emotional lives of its detectives.

Duty vs. Obligation

In the police drama, the interrogation room acts as a space for the enactment of a mostly conservative, duty-based form of social responsibility. However, the interrogations on *NYPD Blue* cover a much wider range than Milch's description of the "devil's bargain" suggest. There are two broad types of confessional strategies that this series uses: "negotiated" and "obligatory" confessions. Taken together, these two forms of confession provide a much richer understanding of how the series represents the role of social responsibility as a marker of citizenship. Ultimately, this broadened approach to responsibility as articulated through the act of confession allows us to connect the melodramatic mode to the concept of citizenship through the interpersonal relationships among the detectives.

The most common form of confession by far on *NYPD Blue* is the "negotiated" confession. In particular, I want to focus on two separate types of negotiation that take place within the context of confession: coercion and consent. Both of these forms of negotiation are predicated on absolute guilt and are designed quite simply to elicit an admission from a suspect. The first of these is what David Milch refers to as part of the "devil's bargain" of policing: the coercion of confession through either the threat of violence or actual physical abuse itself (Schiff 1997: 6). The character of Andy Sipowicz is infamous for his ability to beat a confession out of a suspect, but all the male detectives are seen as able (and willing) to use force if necessary. The limits of this approach are put on display in a humorous but illustrative way in a fourth season episode

("What A Dump"). The normally shy and insecure Det. Greg Medavoy, having eaten too much sugar, suddenly becomes abusive in the interrogation room, smashing a suspect across the head with a telephone book. The tactic works and Medavoy is suddenly empowered, although his partner, James Martinez, is unsure about this newfound tough side, especially when Medavoy crosses the line later in the episode. When Medavoy complains to Martinez that nobody questions Sipowicz if he "tunes up" a suspect, Martinez replies that Sipowicz "gets results." Thus, violence against a suspect is seen as part of an economy of social justice: an investment or risk in the hopes of a larger return in the form of a confession. As Milch suggests, it's an investment that society advocates on the part of the detective, but which it will not guarantee.

The second form of negotiated confession is represented as a matter of consent. The consensual confession is offered as an excuse for action under the false assumption that the confession and excuse combined will mitigate the guilty party's actions. In "Twin Petes" (season five), a drug addict suggests that he suffers from blackouts and seizures and, therefore, isn't responsible for robbing two ATM machines. Similarly, in "NYPD Lou" (season one), a young man named Freddy, who has raped and killed a small boy, comes to the conclusion that he needs "treatment" for his actions and impulses after Sipowicz paves the way for his confession:

SIPOWICZ: I know this has got to be tearing you up inside. But you are gonna feel a whole lot better if you just tell the truth. Did you have sex on him?

FREDDY: Yeah. He opened his eyes and he started crying and then he started screaming. I put my hands on his neck to keep him from screaming. That's when I knew that he was gonna tell. I know I got a problem. I should get treatment, huh?

In "Simone Says," (season two, discussed in the previous section) Mr. Davis admits to abusing his daughter only after Det. Medavoy suggests empathy with him: "I'm a family man. I know how important it is to keep families together. Based on what your daughter has said, I think your coming forward is the best chance at keeping this situation under control." Thinking that his punishment might consist of counseling for his "problem," Mr. Davis confesses. At this point, Medavoy arrests him and Mr. Davis accuses Medavoy of lying to him. "You said if I told the truth I wouldn't have to go to jail. You lied to me!" "I said it was your best shot," Medavoy responds. "You want my opinion? I hope they put you away, you sick son of a bitch."

Each of these suspects places the blame for their crime on their own lack of self-control and they place responsibility for that lack outside of themselves. Other suspects also insist that they are not responsible because they were unable to control their environment. In "Heavin' Can Wait" (season three), a young drug addict admits to killing two young children during an armed robbery, but insists that the gun virtually went off by itself. In "Hammer Time" (season five), a crack-addicted woman brutally murders her own daughter and blames the daughter's own inability to control her bladder for pushing her to murder.

A key trope of the series, in fact, is the importance of maintaining control of one's actions and reactions. Police stories are often narratives about control. The job of the street level uniformed officer is to maintain order and control on their beat – to regulate behavior by intervening in disputes; the job of detectives is to close cases by controlling information, which they do in part by controlling interrogations; the job of the "bosses" (sergeants, lieutenants, captains, and commissioners) is to control their squads. All of this is done in order to maintain some level of social order. But the issue of control in *NYPD Blue* extends beyond this imperative of policing and into the lives of the detectives and officers themselves: at the level of their emotions, their behavior, and even their bodies.[12] The lack of control exhibited by suspects is directly opposed to the efforts on the part of the detectives, especially Sipowicz, to establish and maintain control. If Sipowicz is a character on the path to redemption, the suspects he encounters provide a mirror-image reminder of just why his recovery is so important. The worst thing one can be on *NYPD Blue* is out of control. It is a sign of a "thin and defensive" sense of citizenship that represents a "selfish and instrumentalist attitude" in which "rights are demanded, but no responsibilities accepted" (Faulks 2000: 69–70). This vision of "thin" citizenship is nicely illustrated by the description of a young drug addict by her mother following the girl's arrest on murder charges: "Chantelle's mind always full of envy. She always want everything. She see something on the TV, she want it. She don't want to give. She just want to get. There comes a time when you have to give something up or maybe someday, something come and take away everything you got" ("Heavin' Can Wait").

This idea of sacrifice opens up the possibility of the second type of confession: the "obligatory" confession. As opposed to the negotiated confession, which deals in absolute guilt, locates the detectives on opposite sides of the negotiating table from the suspects, and operates at the level of duty-based social and legal responsibilities, the "obligatory" confession works through empathy and results from a crime that cannot be so easily vilified as a lack of personal control. These crimes are often associated with accidents or emotional states with which the detectives can empathize. Of course, the detectives can use the illusion of empathy as a tool for eliciting a "negotiated" confession, as Det. Medavoy does when he offers a hug to a man who has just killed his father's best friend out of jealousy as a way of gaining his trust ("A Wrenching Experience"). For the most part, however, displays of empathy, when they occur, are genuine. In "Simone Says" (season two), for instance, Sipowicz and Simone try to help a young construction worker from New Jersey, named Paddy, who confesses to killing a mob figure who had been sleeping with his girlfriend, Paula, a model trying to shake her small-town New Jersey roots. The bullets from Paddy's gun also killed an older woman sitting alone in her apartment. The detectives are initially furious about the killing of an innocent bystander and go after Paddy as if this was a mob-related killing. Of course, it turns out to be nothing so extraordinary, just a case of jealousy and revenge. But something about the young man's devotion to his girlfriend despite her infidelity touches the detectives and they offer him a way out: indicate that there was a struggle for the gun and that this was a matter of self-defense. Amazingly, the young man refuses the opportunity. He takes full responsibility for his action and accepts the consequences.

What makes this confession "obligatory", as I have defined the term, is that it is voluntary (he had other options) and it comes as an expression of solidarity with the detectives, with the family of the woman who was killed, and with Paula herself. In fact, when Sipowicz and Simone ask for him to confess, they place it in terms of the family that just lost their mother. Sipowicz says to him: "Paddy, I know you're a decent guy. I think you want to do the right thing and help these people who lost their mother. You gotta know what they're going through. You ready to do the right thing, Paddy?" Paddy then begins to tell the story of how he and Paula fell in love when they were just eleven-years-old, and how he found out she was seeing someone else in New York:

SIPOWICZ: So you waited outside in front of her place and you saw him come out?

PADDY: Yeah, he came out.

SIMONE: Did he see you?

SIPOWICZ: Did he threaten you or anything?

PADDY: Nah, just the way he walked, the prick. Like he'd just taken away my life and felt good about it. So, he sees me and I tell him I want to talk to him. He tells me to kiss his ass. So I tell him he should leave her alone because he probably doesn't even love her.

There are, of course, elements of division that work against the idea of solidarity: clearly Paddy and the mobster share no common ground, except for Paula. And Paula is positioned as a potential object here: treated as a possession. But the confession is framed from the outset as helping out the other family, and Paddy is willing to own up to his responsibility: "I'm sorry about that lady," he tells the detectives after he has told the story of what happened. He is also adamant about his love for Paula, not as an object (something she represents in her role as a fashion model in New York), but as his "life." This devotion to Paula is underscored in their final scene together:

PADDY: One minute to the next, everything's just....gone.

PAULA: I'm so sorry. I'm so sorry for hurting you. Don't hate me, OK? Just try not to hate me.

PADDY: Hate you? I could never hate you. That's just something I could never do. You know that. Don't you know that? You gotta know that.

Paula reaches across the table and throws her arms around Paddy. The camera then cuts to Simone, who has been in the room with them, his face a study in disappointment that Paddy will be going to jail. The narrative arc of this situation constructs solidarity and empathy on three levels: between Paddy and the family through his acceptance of responsibility for their mother's death; between Paddy and Paula through his confession of devotion; and between Paddy and the detectives through his expressions of honest remorse.

As we can see, the basic element of confession in *NYPD Blue*, an element so central to the police drama as a whole, opens up a broad discourse about what constitutes "good" citizenship. To summarize, narrow or "thin" citizenship, emphasizing either individual rights *or* social responsibility, yields a duty-based view of citizenship. Older series like *T.J. Hooker*, *Hunter*, and even *Miami Vice*, as well as newer series such as *CSI*, *Without A Trace*, or *Cold Case*, focused as they are on representing justice within a mostly episodic narrative structure, offer up a thin, and mostly conservative, version of citizenship that emphasizes punishment for individuals who overstep the limits of their rights and ignore their duty to live within the law. Importantly, each of these series mostly foregoes the acts of interrogation and confession, opting instead for a more violent form of justice that effectively closes down any potential for empathy with suspects and criminals. *NYPD Blue* is an example of a thicker version of citizenship that acknowledges and re-inscribes the duties of citizenship (by punishing its criminals) but at the same time underscores the *obligations* of a more "intimate" citizenship that emphasizes the centrality of interpersonal relationships and the goal of a more humane and empathetic community.

The Importance of Letting Go

Nowhere is this vision of "intimate" citizenship more clearly represented than in the interrelationships between the detectives. Of course, as a prime-time ensemble drama, the series delves into the romantic lives of the detectives with some regularity. Rules against dating, marrying, and sleeping with co-workers are regularly abandoned in order to provide viewers with the level of romantic and sexual tension that television drama seems to require. More importantly, though, these characters are deeply committed and caring friends. In this capacity, as much as in their roles in law enforcement, they represent a thick, holistic, and intimate model of citizenship. While the series is frequently criticized for its overly melodramatic tone, it is precisely within those melodramatic moments that the most powerful demonstrations of this intimate citizenship emerge.

As I discussed earlier, one of the key themes of *NYPD Blue* is the importance of control. A corollary to this focus on control, however, is the importance of "letting go" and "reaching out": two themes that are widely present in the series and help to establish a more complex discourse about control. Letting go is constructed in the series as central to maintaining perspective in an occupation that often confronts the detectives and victims with acts and people that they cannot control. In a scene that mirrors the mother's need to confess her culpability in "Lost Israel," another mother tries to take responsibility for her daughter's death at the hands of her ex-husband. "She was the one thing that I did right. And then I let this man into our lives." To this, Simone responds: "No. No. You are NOT responsible for this. You did not know what kind of person this man was. We can't always know what's in another person's heart" ("Large Mouth Bass"). In another episode ("Heavin' Can Wait"), Simone and Sipowicz investigate the murder of a 5-year-old boy and his 7-month-old brother. The sight of these victims obviously shakes both of the detectives and their tempers flare throughout the episode. In an attempt to explain his anger, Sipowicz opines: "There's no helping anybody." Sipowicz 's frustration with the case is exacerbated by two other problems in the episode: his wife's violent morning sickness, and the fact that Russell calls on Simone instead of him, her AA sponsor, when she fears that she is about to start drinking again. The final scene of the episode finds Simone alone on the roof of the station house tending to his pigeons, which he sets loose. This final act calls to mind Peter Brooks' argument in which expressivity represents a "victory over repression" and fixes "in large gestures the meaning of [characters'] relations and existence" (1976: 4). In these terms, the flight of the pigeons works to "articulate all that [the story] has come to be about" (1976: 4): the importance of finding ways to keep the trials of life from consuming you.

It is when the characters fail to find a productive release that they risk being consumed emotionally and physically, particularly by the habits that they fight to control. As I've already indicated, both Sipowicz and Russell are recovering alcoholics. Other primary characters battle different demons, such as overcoming feelings of abandonment and controlling violent impulses. What is important about these behavioral problems is that the detectives embody them differently than the suspects and criminals. In fact, what seems to draw the line between virtue and corruption most clearly in this series is the ability to control one's impulses and to discipline one's behavior – even if you have to ask for help.

As we've seen, criminals on *NYPD Blue* give in to their impulses in dangerous and unrestrained ways. They beat their children to death, or abuse them, or strangle them ("Hammer Time," "Large Mouth Bass," "Lost Israel"). They drink too much and give in to petty jealousy and greed ("Simone Says," "A Wrenching Experience," "Twin Petes"). Their greed puts them in situations where they are either killed, jailed, or commit murder themselves ("Heavin' Can Wait," "Where's 'Swaldo," "Large Mouth Bass"). The detectives, on the other hand, are models of control and responsibility compared to the criminals. They may be violent, but they only resort to violence when it serves a productive purpose, such as getting a confession from someone they know is guilty. They may drink too much as well, but they struggle with their drinking in productive ways and treat each other with compassion in their efforts to help. And, most of all, they know when to let go, surrender control, and to *learn* from mistakes and misfortune.

These characters are put in harm's way, forced to sacrifice their innocence, and then asked to struggle to get it back somehow. As the series tells us over and over again, learning control is, paradoxically, a matter of learning how to let go. After Simone's death in the sixth season, Russell returns to the job only to find herself feeling incapable and unproductive. In one episode ("Show and Tell"), her partner, Jill Kirkendall, visits her at home where Russell is holed up with a bottle of vodka that she bought but did not drink. Knowing that she was in trouble, Russell asked Kirkendall to come help her. "I want to be strong for him [Simone] and I feel so bad. I need help," says Russell. Kirkendall assures her that she will get help. "I want to go to work," says Russell. "If I don't go to work, I'm gonna stop going." The episode establishes the parameters of an intimate citizenship quite clearly: the need to take responsibility for oneself but to also allow others to show empathy. Russell needs to continue her duties as a productive member of the squad, but at the same time she needs to admit that she cannot do it alone.

The dialectic of discipline and letting go that forms this intimate citizenship is repeated in a later scene in which Russell visits Linda Krause, the representative for police widows and orphans, and admits that she is feeling lost and helpless:

KRAUSE: I guarantee you're going to get your husband back. You'll learn ways it won't hurt to remember.
RUSSELL: I feel like I'm losing him.
KRAUSE: You feel like coming in here cost you some of Bobby?
RUSSELL: I feel like I'm losing more and more of him. Everything hurts.
KRAUSE: You gotta let it. That's the only way it's going to get better.

By "learning ways" to cope with her grief, Russell will gradually regain control of her emotions as well as her memory of Simone; but first she must allow herself to experience that grief fully. It is precisely her attempt to control the grief by repressing it that is threatening her from within. Once she can externalize the grief she can gain control of it in more productive ways. The important rule in all of this, as presented by the series, is that she cannot do it alone.

Conclusion

The public controversy over *NYPD Blue*'s forays into edgier language and displays of nudity missed the ways in which the series worked to construct a deeply moral (and moralistic) world. *NYPD Blue* addressed itself to a complex set of ideas about "how to be" in this world. The series did not offer easy answers to its moral questions. Instead, it constructed a version of citizenship in flux – always being formed in relationship to the realities of the moral world in which it functions.

The model citizen in *NYPD Blue* is constructed through the melodramatic mode as an empathic figure who is capable of balancing individual rights with social responsibilities. The series does this, in part, by drawing parallels between the detectives and criminals that highlight the distinctions between their decisions and actions. What marks the detectives as distinct from the suspects (and even the victims), and measures their growth and development as citizens, is their ability to gain and maintain control in the midst of personal and professional chaos.

Citizenship, as represented by the detectives on *NYPD Blue* is not an ideal and finished state. Rather, it is a dynamic and ongoing process of balancing one's duties and obligations. The detective heroes of the series are in a state of constant construction and re-construction. The moral world of *NYPD Blue*, as with police dramas in general, is made legible through the interactions between its villains and heroes. But in keeping with Peter Brook's arguments about melodrama, the point is not the eventual triumph of virtue over corruption, but the establishment of a moral world made legible through action and interaction. The characters on *NYPD Blue* are too complex to be easily positioned as simply "good" or "evil." Instead, their virtue (or lack thereof) is marked by their ability to control their (often misguided) impulses and, paradoxically, to admit that they need help in learning that discipline. Virtue, in the end, is action tempered by humility.

6

ONE THING LEADS TO ANOTHER

Crime and the Commerce of *Law & Order*

Law & Order premiered on NBC on 13 September 1990 and has since become one of the most popular and profitable series in television history.[1] The series enjoyed one of the longest continuous runs in broadcasting history (ending its original run in 2011) and has been reaping the benefits of cable syndication since 1994, when it was first rerun on A&E. Dick Wolf, the creator and executive producer of the series, once described the quality of the writing on the show as "'no fat' writing – writing that tells the story in which each scene flows into the next with the inevitability of falling dominoes" (Wolf 2003: 10). But the inevitability that Wolf describes is not a matter of foregone conclusions within the plot of the episodes. On the contrary, perhaps the only thing truly inevitable about the stories that *Law & Order* offered each week is that the movement from conflict (dead body) to resolution (conviction or acquittal) was riddled with contradictions, complications, and compromises in the way that justice is understood and ultimately served.

Like the flow of the episodes themselves, the series began as one thing and has become quite another. Most simply, and like nearly every series, *Law & Order* was conceived as a strategy for market differentiation – a way to address the waning fortunes of one-hour dramas on U.S. television. Along the way, however, the series became nothing less than a cultural institution and a lesson about the dynamics of the contemporary television landscape. On the one hand, *Law & Order* is the epitome of the new business of television: an expansive brand identity comprised of seemingly endless spin-offs as well as increasingly lucrative syndication deals. This expansive business model has, at the same time, cultivated the seeds of an expansive cultural forum around the notion of crime itself: what counts as crime and what to do about it. This forum operates on three levels: within the police genre as a whole, within the series itself (within and between episodes), and within the franchise (between different series that carry the *Law & Order* logo). Holding these two notions in balance, I will argue in this chapter that *Law & Order* offers us a fruitful opportunity to consider the intersections between commodities and the culture in which they circulate – what Eileen Meehan has called the relationship between the culture industries and industrial culture (1986: 564). In short, *Law & Order* usefully illustrates the important connections between the economics of the industry and the narratives that they inspire and, more centrally, that the very notion of television as a cultural forum must be grounded in the specifics of economic necessity and opportunity.

Understanding *Law & Order*, then, requires that we hold economics and aesthetics in careful balance. In what follows, I will address the three primary avenues through which the interplay between commerce and culture in *Law & Order* may be most fully understood: storytelling, syndication, and spin-offs. Within these broader sections, I will focus on discursive potentials opened up by the focus on story over character (in particular, how the story-driven narratives enter in discursive struggles over what constitutes crime), the commercial potentials in syndication opened up by these story-driven narratives, and the ways in which the franchise that has been built around the *Law & Order* brand constitutes a forum in itself by expanding the approaches to crime and punishment across the brand. Though the various series within the franchise, taken together, resort ultimately to a vision of crime as the responsibility of individuals (rather than social structure per se), *Law & Order* has done more than any other series to perform the discursive and dialogic functions of the police drama that has been the primary focus of this project.

What Was (and Is) *Law & Order*?: Series Structure and Syndication

Unlike its closest brethren of the 1990s – *Homicide* and *NYPD Blue* – *Law & Order* was not a serial drama. Rather, the series was decidedly episodic in structure and followed a strict formula. Each episode contained a single case, often "ripped from the headlines,"[2] which was then broken down into two half-hour segments: the first half-hour told the story of the police investigation which inevitably leads to an arrest; the second half-hour was the story of the criminal prosecution of the suspect by the District Attorney's office that did not always result in a conviction. The cast contained only six regular characters at any given time (two police detectives, their Lieutenant, two attorneys in the DA's office, and the District Attorney him or herself), and the cast experienced a complete turnover since the first season. Despite this relatively simple and rigid structure, the episodes were designed to allow for an extraordinary range of voices as the principal characters explored the vagaries of the case from a number of competing angles. As Dick Wolf, the creator and executive producer of the series is fond of saying: "the 'perfect' episode is one where all six characters have different points of view on the same moral conundrum or idea and they're all right" (Longworth 2000: 14).

The series flew in the face of the conventional programming wisdom of the time in three respects. First, the show was conceived as twin half-hours that could potentially be broken apart in syndication. According to Wolf, the main impetus for the split episodic structure had to do less with aesthetics than with economics, specifically the syndication market. "In 1988, there was no aftermarket for one-hour shows" (Svetkey 1999: 30). In fact, the networks seemed to be moving away from hour-long dramas in prime-time, opting instead for cheaper (and often more popular) situation comedies, news magazines, and reality shows in an effort to cut costs and protect themselves from labor disputes and costly contract negotiations. But news magazines and reality shows have only a limited "shelf-life" at best and are too quickly outdated for any real value in syndication (Littleton 1995). Thus, despite the potential burden that hour-dramas place on network budgets, they do have more long-term value than most of the cheaper reality-based programming.[3] This extended market value for dramas has risen with the increasing competitive presence of cable networks like USA, fX, TNT, Lifetime, and A&E, all of which have profited greatly by stripping one-hour dramas.

The second way that *Law & Order* broke with the programming logic of the day was through its closed narrative structure. In the late 1980s, when *Law & Order* was first imagined, the prevailing wisdom about drama was that audiences wanted characters that they could get to know. But the writers of *Law & Order* took a different approach – one that harkened back to the days of *Dragnet*: a narrative world that "is indifferent to its characters' personal lives...except as they affect peoples' jobs" (Sam Waterston, quoted in Svetkey 1999: 31). Of course, information about the characters' private lives *was* revealed over the course of the series; but as Wolf himself points out, it came to light within the context of interpersonal relationships rather than through misplaced personal disclosures for the sake of building characters quickly. "There's constant character revelations in human relationships, and I think that in *Law & Order* it is actually parceled out much closer to the reality of day-to-day life" (Longworth 2000: 11).

The lack of attention given to the private lives of the show's characters leads to the third way that *Law & Order* challenged traditional programming logic: regular cast changes. While ensemble casts typically do undergo some change over the course of a series' run, no series had discarded and added characters with as much regularity and success as *Law & Order*. The impulses to replace characters run the gamut from network demands for different types of audience appeal, to actors' desires to move on, and the producers' desires to change gears as well. Explaining the ease with which the series has weathered cast changes, Wolf suggests that the real stars of the series are its writers, not its name actors. And while programmers and producers often worry that the loss of a major star might negatively impact the ratings for their series, the variety in the cast became one of the "enduring hallmarks" of the series, perhaps even one of its principle attractions (Auster 2000: B18).

The combination of a rigid format and a malleable casting arrangement explains a great deal about why *Law & Order* has experienced almost unparalleled success in its off-network life, which began in 1994 when A&E purchased the rights to the first 181 episodes of the series for $155,000 per episode. *Law & Order*'s run on A&E was so successful that it's price nearly doubled in 1999 when TNT bought the rights to these same episodes for roughly $250,000 per episode (Katz 1999: 1).[4] As mentioned earlier, the original idea for *Law & Order* stemmed from almost purely economic considerations: an interest in half-hour dramas at the networks and, more centrally, a related weak syndication market for one-hour dramas. Dick Wolf's idea to split the narrative into two half-hours was an effort to address these conditions and try to safeguard against failure. The one-hour episodes of *Law & Order* would be comprised of two half-hour segments that could be broken apart later if necessary. Wolf wanted to have the flexibility of the half-hour length in syndication but also be able to deliver network programmers something that they could use:

> The problem with doing half-hour drama is "What do you do with the other half-hour?" You don't want to program a sitcom and a drama in the same block, so you're forced to find a mate. And out of that grew the idea of doing a one-hour show which could be split in two – to give two discrete stories in the same hour.
>
> *(Dick Wolf quoted in Lindheim & Blum 1991:127)*

This split narrative structure would provide network programmers, off-network syndicators, and Wolf's production company with the maximum amount of flexibility within the current programming market.

While the basic idea of a split structure is evident in *Law & Order*, the degree of differentiation that Wolf had imagined initially never became a reality in the episodes that he finally produced for NBC. The details from the police investigation (and often the police themselves) have a necessary function in the second half of the episodes. To put it another way: though Wolf has frequently underscored the split format by proclaiming that "the first half of the show is a murder mystery; the second half is a moral mystery" (Carter 1997: C11), the moral mystery is impossible without the details of the murder mystery. Thus, instead of providing a series that could be split neatly in two stand-alone half-hours, Wolf provided a one-hour format, comprised of legal and procedural elements (kept *mostly* separate from one another) and a closed, story-driven plot, that allowed for enormous flexibility within the structure of each episode. "The wonderful thing about the format," Wolf told two early interviewers, "is that it frees you up to do almost anything. For example, there may easily be a show where the prosecutors are in the first half" (Lindheim & Blum 1991: 133).

Given that Wolf dropped his initial strategy for securing success in the syndication market, how can we account for the success that the series *has* enjoyed in the aftermarket? The answer rests in the way that the story-driven plot (rather than the split format) addressed the demands of the syndication market. Because the episodes themselves do not illustrate the clean split between half-hours as clearly as Wolf's early proclamations would suggest, we can assume that the market "problem" *Law & Order* addressed in terms of programming was not the one-hour dramatic format per se, but the *serialized* dramas that had become such critical successes during the 1980s. Rather than the half-hour sections (which never came to full fruition), it is the presence of the story-driven narrative *within a changing syndication market* that has led to syndication success for *Law & Order*. More importantly, the "innovations" that *Law & Order* brought to the police genre – less character, more story – were a product of thinking about the long-term economics of the television industry. In other words, *Law & Order* highlights the degree to which textual and discursive innovations within a genre and/or cycle are forged within the dynamics of the market for certain kinds of products that can ensure long-term profitability for producers and programmers alike.

The syndication and off-network market for one-hour dramas in 1988 (when *Law & Order* was initially conceived) was in a state of flux that was complicated by at least three factors: the somewhat contradictory prospects for one-hour dramas; changing economics in the face of deregulation; and the emerging fortunes of the cable industry. Taken together, these factors help to explain both the impulse to create *Law & Order* as a particular kind of narrative, as well as the value that *Law & Order* eventually came to represent for both cable and broadcast networks hoping to build their own brands. Before engaging with *Law & Order* directly, it will be helpful to establish the historical contexts for this changing syndication market for one-hour dramas and the emergence of cable as a viable location for off-network programming.

By 1988 it was clear to programmers and producers that one-hour serial dramas were not as valuable in syndication as they had hoped. Perennial ratings powerhouses like *Dallas*, *Hill Street Blues*, and *St. Elsewhere* had entered the syndication and off-network market only to see their value decline rapidly as their second-run ratings (and then clearances) dropped. Even as *Dallas* sat near the top of the prime-time ratings in the 1984–5 season, its value in syndication was in question. In 112 syndication markets, *Dallas* had captured only one-third of the audience that *Wheel of Fortune* was getting. But at the same time, many of the stations were happy with the quality of the ratings, especially among 18–49 year-old women (Kaplan 1987: 1). Three years later, however, both *Dallas* and *Falcon Crest* were performing poorly in the syndication market, so much so that by the 1987–88 season both series requested that they be allowed to drop their serial formats in order to "improve their chances in the soft syndication market for one-hour shows" (Mahler 1988: 83). Similarly, *Hill Street Blues* and *St. Elsewhere*, both produced by MTM, also faced difficult syndication futures. *Hill Street Blues* was purchased in most of the top markets in 1987–88, and was relegated to late-night slots where it was often quickly replaced. *St. Elsewhere* performed so badly in syndication (purchased by only 20–25 stations in 1988) many speculated that its low repeat value led ultimately to its cancellation by NBC in 1988 (Mahoney and Mahler 1988: 3).

Although the syndication market seemed to have gone soft for one-hour dramas in general, it was the serialized dramas that were having the most difficulty. As CBS president Kim LeMasters stated in 1988, the serial format "just kills you when you go into repeats" (Mahler 1988: 83). Episodic one-hour dramas, on the other hand, were performing quite well in syndication. *In the Heat of the Night* cleared 153 markets with 87 percent national coverage via syndication on independent broadcast stations ("MGM Domestic Television"). Meanwhile, MCA, the studio that produced *Murder, She Wrote*, sold the drama to the USA cable network for $300,000 per episode – the highest price paid by a cable network at that time for a single series. What *In the Heat of the Night* and *Murder, She Wrote* have in common is a closed, episodic narrative structure and solid prime-time ratings. Thus, while the syndication market for one-hour dramas had mostly softened in the late 1980s, episodic dramas demonstrated that there was still value in the form, especially episodic narratives that offered clear resolutions each episode.

The difficulty in syndicating – or "stripping" – serialized one-hour dramas was directly related to the power of independent stations in the early 1980s. Between 1980 and 1986, the number of independent stations in the U.S. increased from 108 to 249 and the syndication market experienced a subsequent boom (Boyer 1986: 26). These stations, however, were more interested in the scheduling flexibility of half-hour sitcoms than more expensive and less flexible one-hour dramas, especially those that were serialized, which made them even more inflexible in terms of scheduling. Additionally, sitcoms have a longer life in syndication – typically able to support five to six repeats, while dramas lose significant audience share in repeats, making them far less attractive to independent stations that rely on multiple repeats of individual episodes (Boyer 1986: 26). Given the reduced demand for one-hour dramas, studios were forced to either cut costs (because of lower network license fees) or move away from the one-hour format altogether. Looking back to Dick Wolf's initial idea for *Law & Order*, the economic imperatives for the split half-hours and the (cheaper) formulaic structure become clear: both are connected to the syndication market. The split half-hours would make the series more palatable for independent stations, and the formulaic structure could potentially ease production costs and reduce the deficits that the producer would have to make up in syndication.

When the growth of independent stations leveled off in the late-1980s, the syndications market also flattened out. The initial growth had opened the market to a great deal of product and had dictated the kinds of programming that producers were willing to risk – if independent stations wanted sitcoms, that is what producers would give them. As the growth of independent stations slowed, however, the significance of cable networks grew. In 1988, cable penetration in the

U.S. had reached 50 percent (up from 10 percent in 1972), and between 1981 and 1985 the cable industry experienced a second wave of growth that included the emergence of networks such as Disney, Lifetime, Playboy, FNN, the Weather Channel, Discovery, TNN, A&E, USA, and of course, MTV. These new outlets demanded programming and were willing to pick up series that they could purchase cheaply, that would fill their schedules, that might offer them prestige, and that would support their brand identity. Thus, for example, Lifetime was able to acquire *Cagney & Lacey* in 1988 for approximately $100,000 per episode, as compared to the $1.7 million per episode that *Magnum P.I.* was able to command in the broadcast market in the same year (Ivey 1988: 36).

The steady rise in cable penetration and the increasing value and presence of key cable networks led regulators to reconsider one of the central components of the syndication market: the financial interest and syndication rules – or "fin-syn" – established in 1972. Under fin-syn, major broadcast networks could no longer own a significant portion of the programs that they scheduled. This rule was designed to encourage and protect production companies, and provide them with a fair economic stake in the syndication aftermarket for their series. In short, under fin-syn, production companies would produce their series at a deficit during its initial run, receiving only a fraction of the total production cost from the networks in the form of license fees. In exchange the production companies would retain ownership of their series, which they could eventually sell in the syndication market. If a series was a hit during its initial network run, the production company stood to make significant profits in syndication by selling their series in each local television market for a specified period of time.

Efforts toward repealing fin-syn were predicated on the argument that the lack of competition for programming, which was the initial impetus for the rules, was no longer an issue. Cable networks were steadily eating into the audience that broadcasters had once commanded. In the face of declining advertising revenues, the broadcast networks applied pressure to an FCC that was already geared toward deregulation of the industry. The FCC slowly loosened the restrictions beginning in the late 1980s and finally repealed the fin-syn rules in 1995 (McAllister 1997).

The effects of this repeal on the market for one-hour dramas were almost immediate. One effect was increased pressure applied to producers by broadcast networks in terms of financing and the "back end" (money that is made from syndication and off-network licensing). As networks saw their share of the audience drop from 90 percent in the 1970s to 65 percent by 1990, their willingness to pay increased license fees for expensive one-hour dramas declined; as producers were forced to pay more on the front end (and accrue massive debt in the process), their return on the back end (the "prize" of fin-syn) suffered as well. A second effect, related directly to this first effect, was that several producers opted to take their dramas to cable. For instance, producers like Steven J. Cannell (*Street Justice*), LBS/All-American (*Baywatch*), and Paramount Television (*Star Trek: The Next Generation*) each declined the "network-dictated licensing arena in favor of what distributors claim is a financially and creatively less restrictive free-market approach in syndication" (Freeman 1992: 22). By selling their series in first-run syndication and bartering for advertising space, these producers could fully finance their series and turn a profit from the beginning. Despite declining fortunes, however, the major broadcast networks still command the largest audience for their series. This continued dominance explains why not all producers are taking a similar route to first-run syndication on cable.

Most expensive dramas like *Law & Order*, in fact, still begin life as part of a major broadcast network schedule and this broadcast affiliation establishes the limits of how the series will function in the aftermarket. Additionally, as cable networks, producers, and broadcast networks are increasingly clustered under the same corporate umbrellas, the movement of a series from broadcast to syndication no longer represents a movement from one economic structure to another as it did under fin-syn. With the demise of fin-syn and the concomitant restructuring of the industry around vertically integrated conglomerates, any explanation of the function and value of a series in syndication needs to take into account the intertwined fortunes of producers, cable networks, and broadcast networks.

First, and most obviously, the off-network market offers significant opportunities for producers in the form of increased revenue. Under the fin-syn rules, once a series went into syndication, the producers of the series stood to make windfall profits, provided that the series was popular in repeats. Whether these producers sell the rights to their series for cash only, or barter for advertising space, profits from syndication are often in the hundreds of millions. Another potential benefit of entering the off-network market while the original series is still on a broadcast network schedule is expanding the audience for the series as a whole, which makes the property more valuable in the long run. As mentioned earlier, A&E initially paid only $150,000 per episode for *Law & Order*. This relatively low price was the product of a soft syndication market for one-hour series on independent stations and affiliates along with the series' relatively low ratings (although, like *Hill Street Blues* before it, it was most popular in cable households). By the time that the off-network rights for *Law & Order* came up for renewal in 1999, TNT outbid A&E and USA for the rights to the series, paying nearly twice as much ($250,000) for the rights to the 181 A&E episodes and then $700,000 for each subsequent episode – a deal that industry analysts suggested might be worth up to $150 million dollars (Katz 1999: 1).

Dick Wolf and MCA-TV were not the only beneficiaries of the A&E deal; NBC benefited as well in the form of vastly increased ratings for original episodes. The ratings for the first seasons of *Law & Order* were lean – the series failed to finish in the top thirty during its first four seasons. As discussed earlier, the series was threatened with cancellation if it didn't get its numbers up by including more female cast members. While cast changes may have had an impact on the ratings, a more direct (and surprising) impact came from the off-network run on A&E.

In 1994–95, the first season following the series' appearance on A&E, *Law & Order* broke into the top thirty in the ratings; by the end of the 1998–99 season, it was nearing the top ten (Brooks and Marsh 1999: 1258–1259). Though popular wisdom held that having a given series on both a broadcast network and a cable network at the same time might hurt ratings at both networks, Dick Wolf calls the practice "totally synergistic" (Richmond 1998: 108). Michael Cascio, vice president of programming at A&E in 1998, explained this synergy in terms of cultural distinctions between audiences for cable networks like A&E and those for traditional broadcast networks: "A&E has introduced new people to *Law & Order* who have never watched it on NBC" (Richmond 1998: 108).

Finally, A&E (and later TNT) accrued significant benefits from *Law & Order*. Most obviously, the presence of a quality drama on a cable network offers clear opportunities for building market value through increased ratings in important demographic groups. By 1994, the year that *Law & Order* began its first off-network run on A&E, cable had clearly established its value, reaching 64 percent penetration. Cable networks like A&E had conquered their initial hurdles of simply finding enough programming to capture an audience and were now able to begin building their brands in new directions. In 1993, for instance, the audience for A&E increased by 32 percent, including an 18 percent increase in primetime and a 57 percent increase in daytime (Hettrick 1994). These increases in audience (and, by extension, advertising revenue) allowed A&E to bid seriously for more expensive and prestigious off-network series like *Law & Order* and to prove to producers that they were "a viable customer for off network product." A&E's purchase of *Law & Order* continued to build that value and "quality" identity. Brooke Bailey Johnson, the senior vice-president of programming and production at A&E in 1994, highlighted the fact that *Law & Order* was "an appropriate show for us, and there are not many that are." What made the show appropriate, according to Johnson, was the combination of high production values and intellectual engagement. Perhaps more importantly, the show was appropriate and important for the network because, according to Shelley Schwab, president of MCA TV, it "put A&E on a new plateau" (Walley and Tryer 1994: 3).

But perhaps what made the show most appropriate for A&E (and later TNT) was the combination of quality and flexibility that it offered programmers. While broadcast networks can profitably exploit serial narratives by highlighting the "event" status of individual episodes and cultivating "appointment" viewing through the continuation of narrative threads, off-network programming presents different constraints and opportunities. Daily stripping outside of prime-time favors episodic narratives. As Michael Cascio suggested in 1998, the reason that *Law & Order* has performed so well in syndication – to the point of bringing new viewers to original episodes on NBC – is that it is not serialized. "You're treated to a complete story in an hour. That lends itself better to cable than shows that are serialized day to day and week to week" (Richmond 1998: 108). In the cable landscape, where programmers shuffle schedules more frequently, where one-hour dramas are as likely to be scheduled in the daytime as they are in primetime, and where they are most often stripped daily, episodic narratives provide much more flexibility for the networks and viewers alike.

The initial impulse to create *Law & Order* with an eye toward the syndication market highlights the importance of engaging with the multiple sources and directions of influence in television production. Rather than simply reacting to and choosing from what is available in syndication, independent broadcast stations and large cable networks can influence network programming in profound ways by creating specific market conditions within which producers must exist. When independent broadcast stations and local network affiliates were the primary syndication market, their disinterest in one-hour dramas drove production companies and networks toward a concentration on half-hour series – mostly situation comedies – that these stations were eager to program in order to fill their available programming slots most profitably.[5] Conversely, as cable networks emerged and matured in the 1980s and 1990s, their drive for prestige and quality niche audiences favored one-hour dramas. But at the same time, they favored the scheduling flexibility of episodic narratives. Thus, the episodic, self contained, story-driven narratives of *Law & Order* are *both* an aesthetic reaction to current programming trends within the police genre (a way to differentiate it as a generic text) *and* a strategic economic decision designed to respond to and exploit existing trends in the syndication market.

The Story-Driven Narratives of *Law & Order*

Dick Wolf and his staff of writers, in their approach to *Law & Order*, made a decided move away from a near obligatory focus on character and the emotional toll that crime takes on police and lawyers in favor of story-driven plots. The reasons for this focus on story over character can be explained from three perspectives. The first is through the aesthetics of realism as a point of differentiation from other dramatic series on television. As Dick Wolf points out somewhat sarcastically:

> The one thing we do on *Law & Order* is the character information is doled out much more realistically than it is in most shows. The term I've used is that it's dispensed in an eyedropper as opposed to a soup ladle. And you look at a lot of pilots and you know when the main characters were toilet trained, people sort of regurgitating their inner secrets and their souls on camera, which makes my teeth itch. But if you think back to the way you get to know most people, they don't stand up in the first three hours of a business or social relationship and tell you everything about their parents, their siblings, where they went to high school, and when they lost their virginity.
>
> *(Longworth 2000: 11)*

Slowing down the speed with which personal information is disclosed, then, is firstly a matter of crafting a more realistic sense of human relationship development. The impulse away from detailed disclosure does not mean, however, that the characters are not imagined from the outset as fully formed and complex human beings. Consider the description of Sgt. Max Greevey as written in the pilot script for the series:

> He's been married to his high school sweetheart for twenty-six years and has three daughters, nineteen, fifteen and six, all of whom can wrap him around their little fingers. He's also an NBA fanatic, and a Knicks fan, an unfortunate combination that has cost him thousands in lost bets since the team's glory days of the early seventies. Greevey's been a cop for eighteen years and has seen everything twice, but he can't stand seeing the bad guys win.
>
> *(Lindheim & Blum 1991: 257)*

The backstories of other characters are crafted with similar specificity, a fact born out in the detailed descriptions that Wolf provides in his companion book to the series, *Law & Order: Crime Scenes*. With regard to Det. Mike Logan, Greevey's partner in the first season, for instance, Wolf writes: "In his backstory, I made Mike Logan the son of physically abusive, alcoholic parents who had raised him as a strict Catholic and used religion as a justification for punishment" (2003: 116). From the perspective of production, these backstories, while secondary to the flow of the plots of each episode, are established early on for the sake of continuity in the writing. This specificity of character also recognizes that audiences do desire characters with whom they can identify. From a narrative perspective, the basic structures of these backstories provide the foundation for the conflict that drives much of the drama in the series: a point to which I will return shortly.

Another aesthetic impulse away from character development is the tempo of the plots themselves. The detectives of *Law & Order* are ruthless in their movement from point to point during their investigation. And the lawyers are equally focused and efficient even in their potentially ponderous discussions of legal ethics and the nature of justice. A typical episode of the series contains an average of between eleven and thirteen scenes over four acts, whereas other prime-time dramas contain five or six (Littlefield 1996: 5). As producer, Rene Balcer argued, "We're trying to cram 20 pounds of show into a 10-pound bag, so there isn't room for anything else" (Svetkey 1999: 29). The primary reason for this kind of "story acceleration" is the split format of the series. As Dick Wolf says, "the reality is that *Law & Order* is a show that if you put in establishing shots and drive-ups and people talking more about their personal lives, *either* half of it could be a perfectly acceptable hour show" (Longworth 2000: 12). The split format of the series, however, demanded that the writers embrace an extraordinarily economic style in order to simply squeeze all of the details of the plot into forty-six minutes of screen time.

Wolf attributes the spare style of the series to two primary influences. The first of these influences is his advertising background.[6] "Advertising is the art of the tiny. You have to tell a complete story and deliver a complete message in a very encapsulated form. It disciplines you to cut away the extraneous information" (Ross 2000: 18). The second influence is his early interest in the Sherlock Holmes stories of Sir Arthur Conan Doyle. Wolf locates the "rigid storytelling, the construction of facts as the thing that first captivated him ... 'That had a seminal effect on the way I saw writing and storytelling. If you can set a character in a story that is compelling and has a backbone, you draw people in'" (Ross 2000: 18). What keeps the story compelling for Wolf is, at least in part, the tempo: "You get bored when it moves too slow. It's amazing what people absorb if you feed them meat instead of filo" (Longworth 2000: 12).

In addition to these aesthetic reasons, the second perspective from which one might understand the focus on story over character is economic. If *Law & Order* is not character driven, it is also not *star driven*. But CBS's criticism of the series as having "no breakout stars" may, in fact, be one of the defining characteristics of its success and longevity. As one reviewer noted, the tempo of the plot "never allows any one of its ensemble cast to break out too big" (Littlefield 1996: 5). While the lack of a breakout star figure might cause network executives to worry over immediate returns on investment (in the form of audiences who tune in to see the star), this same absence has positive economic consequences as well. Just two years before Warner Brothers extracted a record sum from NBC of $13 million per episode for the rights to *ER*, Dick Wolf noted that *Law & Order* was still operating on a relatively efficient budget. "In our seventh year it costs about $1.5 million an episode. You don't get a show hanging on for that long if you give in to hyper-inflated salary requests" (Littlefield 1996: 5).

One of the keys to maintaining this efficiency in production cost was to alter the cast on a fairly regular basis so that no one actor became so indispensable that he/she could begin to dictate the economics of the series. But this "revolving door" of characters was not a part of the initial design for the series. Rather, it was a fortunate accident driven solely by the demands of the network to increase the series' ratings. The first cast changes came after the third season (1992–93) when NBC demanded that Wolf bring some female cast members into what was considered "the most testosterone-driven show on television" (Carter 1997: C11). Warren Littlefield informed Wolf that "virtually no women were watching, and that if I didn't put women in the show it couldn't survive" (Svetkey 1999: 27). Wolf responded by firing Dan Florek (Capt. Donald Cragen) and Richard Brooks (ADA Paul Robinette) and replaced them with Jill Hennessy (ADA Claire Kincaid) and S. Epatha Merkerson (Lt. Anita Van Buren), respectively. These moves paid off almost immediately as the rating/share average for total households during the 1993–94 season had climbed from a 10.2 rating/18 share in the

1992–93 season to 11.6 rating/20 share. Furthermore, the ratings for women 18–34 climbed from 4.9 to 5.8, and the ratings for women 18–49 climbed from 6.3 to 7.1 (Tyrer 1994: 40).

It was only after being forced to make this change that Wolf and his writers began to realize the narrative potentials of new characters. Though some viewers were initially concerned that NBC and Wolf were tampering unnecessarily with a series that was working, the addition of the two female characters helped open up the storytelling. In hindsight, Wolf saw NBC's ultimatum as good fortune, "not only because it afforded new opportunities for the writers, but also because it brought a broader sensibility to *Law & Order*." For instance, in the character of Claire Kincaid, Wolf created the series' first "staunch feminist – and the most politically left character to ever appear on *Law & Order*" (Wolf 2003: 136). If the story is the central guiding principle of the series, then the characters themselves can be relatively interchangeable; what matters is how they position themselves in relation to the crime in question.

Thus, what is at stake is not a matter of removing all human quality from the characters in favor of "just the facts." On the contrary, the human qualities of *Law & Order*'s characters are central to the real drama of the series. Wolf insists, in fact, that the focus on story is not designed to eliminate the individual nuances of character but merely to handle them in a more realistic fashion. "I honestly believe that there is not going to be a lack of personality in these characters. But it is not the aim of the show to give them scenes during which they can expound on what wonderful, or troubled, or tortured, or altruistic people they are" (Lindheim & Blum 1991: 134). Nor is the drama of the series located in the breathless apprehension and conviction of a suspect; it is located in discursive struggles about what constitutes crime in the first place (from the detectives deciding to arrest a suspect to the DAs deciding to indict). And since the NBC mandate, each change in cast members was seen in this same light: what fresh perspectives could each new character bring that would provide opportunities for conflict?

This expanded opportunity for conflict is the third means by which we can better understand the importance of the story-driven narrative over obligatory character development in *Law & Order*; and it is where the two strands most clearly overlap. Conflict, of course, was built into the series from the beginning in the form of generational conflict between the principle characters, especially between the police detectives. This conflict serves two purposes. First, it provides opportunities for the kind of subtle, ongoing character development that Wolf describes. Second, it is necessarily folded into the specific case at hand so that it both pushes the plot further along its "inevitable" trajectory and provides a forum in which to engage with the discursive complexity that the case contains. Importantly, the range of positions here is usually multiplied through non-recurring characters specific to individual episodes (suspects, witnesses, defense attorneys) who provide counter arguments to the Assistant District Attorneys.

A scene from the twelfth episode of the first season, "Life Choice," provides a useful example of how conflict is instigated between principle characters and how it is then deflected onto the more episode-specific characters in order to highlight the discursive complexity of the crime.

This episode featured the bombing of an abortion clinic and the search for the responsible party. After interviewing several potential witnesses, including Rose Schwimmer, the leader of a pro-life group called Women For Life, Greevey and Logan walk down the street and have the following exchange:

GREEVEY: Watch. In five years they'll legalize drugs.
LOGAN: Maybe. Maybe not. It's a great country, Max. You can do what you want, when you want, with who you want. Which means if a woman want an abortion she can get one.
GREEVEY: Just because she was careless enough to get knocked up in the first place?
LOGAN: What are you against: abortions or sex?
GREEVEY: Abortions. I like sex.
LOGAN: So, a seventeen year-old should just louse up her whole life by having a kid.
GREEVEY: A seventeen year-old shouldn't be doing what makes babies.
LOGAN: And crooks shouldn't have guns. Get real. Come on, what they should be is more careful.
GREEVEY: Kids aren't careful. That's why they're kids.
LOGAN: Mrs. Schwimmer didn't mention that she had tea with "Miss Toaster Oven."
GREEVEY: I get it. Now every pro-life protester is a bomber.
LOGAN: You agree with them, don't you?
GREEVEY: OK, Don Juan. Any of your girlfriends ever have an abortion?
LOGAN: One ex-girlfriend, who's now married with two kids, neither of them mine.

Following Logan's answer, Max remains silent for an extra beat and the scene ends. It would seem that the writers want to favor Logan's more youthful and pragmatic perspective over his partner's more traditional views.

But rather than get caught up in the intricate nuances of these two detectives' personal views, the episode forcefully returns to the investigations at hand during a meeting with another possible witness, Celeste McClure, another member of Women For Life. McClure asks Greevey if he would want his daughter or his wife to have an abortion. Greevey responds that he would not, but that he wouldn't want them throwing bombs either. After McClure leaves, Logan pushes Greevey further on the subject:

LOGAN: Hey, you know that's a good question she asked about your daughter. Suppose some boyfriend knocked her up…

GREEVEY: Look, we are not investigating my daughter, my sex life with my wife, or my opinions on contraception! We are investigating the bombing of an abortion clinic! Let's just stick to the job. All right?

This time, it is Logan that is left standing silent as the scene ends. Importantly, Greevey's position is never fully articulated. Instead, it is pushed aside in favor of the job at hand, suggesting that the writers are advocating a pro-choice stance more in line with Logan's earlier perspective. As I will point out, though, the trajectory of this scene establishes the writers' unwillingness to take any particular stand on the issue of abortion itself.

Once the detectives turn up evidence (in the form of a bag of ammonium nitrate) that links both Celeste McClure and Rose Schwimmer to the bombing, the episode then turns to the legal side and the task of establishing the clear guilt required for a conviction. This legal narrative opens up even more questions and conflicts than the investigation itself, which comes to a close when enough evidence has been compiled to simply link someone to some particular act. The question for the legal side of the series is: who is ultimately responsible for the guilty act? Dawn Keetley has argued that the conflict here is between *actus reus* (the guilty act) and *mens rea* (the guilty mind). These two elements of crime "become a matter to be determined and not a matter already determined." Both "are rendered highly visible sites of conflict and are not left as the invisible foundations of easy attributions of blame and of simple answers to the problem of crime" (Keetley 1998: 36). For Keetley, the value of *Law & Order* is its ability to at least suggest that "what constitutes a crime is contingent primarily on the potentially infinite array of lawyers' moral perspectives" (Keetley 1998: 48). In the case of "Life Choice," the process of putting Rose Schwimmer on trial for murder (though she did not actually plant the bomb in question) opens up questions about the distinction between action and responsibility and also opens up the possibility that larger social and political structures may, in fact, have some culpability in individual crimes.

While Keetley goes on to argue that, despite this potential, *Law & Order* ultimately comes down on the side of individual responsibility as the answer – the idea that *someone* has to atone for the guilty act – what remains central to my own argument is the fact that the questions do get raised. As I have stated earlier in this project, the relatively "conservative" visions of justice that most crime shows offer up through individualized solutions and explanations, while quite often troubling, do not short circuit the questions that have been raised and the conflicts that have been aired. Even the "Life Choice" episode, while showing that Rose Schwimmer is clearly guilty of ordering the bombing of the abortion clinic, refuses to take a particular stand on abortion itself – only the act of throwing bombs. The cultural/political dialog on the issue of abortion itself is left relatively open.

This dialog is not contained by a single episode but, rather, can be articulated across the series as a whole, perhaps even the entire franchise. For example, while

"Life Choices" was an episode from the first season of the series, another episode dealing with violence against those who provide abortions appeared in season twenty. The episode, "Dignity," concerned the murder of a late-term abortion doctor (based loosely on the killing of Dr. George Tiller in 2009) and the challenge was to find a way to structure the story that didn't simply reiterate the debates about *culpability* that animated the earlier episode. What the writers focused on was what had changed in the twenty years between the two stories. One of the writers for that later episode, Julie Martin, commented that the first season episode was very much in the writers' minds as they prepared their own script. "The idea that changing medical science, and how that affected *viability* would be the main issue we explored was there from the beginning, since that was what was new about the abortion debate" (Martin 2010: personal interview). Having the medical issue of viability animate the episode opened up a range of new questions that moved beyond the simple pro-choice/pro-life divide and opened up questions of what constitutes a reasonable abortion. A key scene from the episode involves the courtroom testimony of a woman, Lisa Barnett, who had been told by her doctor that her 6-month-old fetus had a severe genetic disorder that was "incompatible with life." Asked if she wanted a late-term abortion, the woman declined, carried the baby to term, gave birth, and held her newborn daughter as she died twenty-one hours later. All of this is revealed in a brave testimony on the stand, moving several members of the jury to tears. During cross-examination, the Assistant District Attorney, Michael Cutter, asks two simple questions:

CUTTER: You're a brave woman, Mrs. Barnett. Is it possible that at another time in your life you might have made a different choice?
BARNETT: Yes, it's possible.
CUTTER: And would you consider a doctor who offers late-term abortions to women in your position as providing an essential medical service?
BARNETT: I would have to say yes.
CUTTER: Nothing further. Thank you.

Two things are important to note about this exchange. First, in terms of style, the scene is played slowly and is constructed to demonstrate the empathy of the ADA and his assistant through a range of reaction shots to Barnett's testimony. The score on the soundtrack, with its string arrangement and minor key, emphasizes the anguish behind the answers. In terms of the plot, Barnett's emotional testimony is initially designed to underscore the compromised mental condition of the accused, suggesting that, having heard Barnett tell this story on a talk show earlier, the defendant was driven to murder. But this legal tactic gets pushed aside during the cross examination in favor of two answers: one social (she might have chosen differently), the other medical (the doctor provides a valuable service). But even though the dialog emphasizes the legal/medical discourses of choice and viability, the scene still contains the residue of Barnett's moving story of her daughter's "dignified" death (an illustration of dignity that is difficult to deny). As with most examples from the series, this episode works hard to avoid taking a final moral position, even if it ultimately takes a legal one.

To sum up this section, the story-driven narrative, which has become one of the defining characteristics of *Law & Order*, needs to be understood in terms of the intersection of narrative and industrial economy. What began as a point of stylistic differentiation – the idea for a split-narrative structure – within a down market for dramas on network television has opened the door to a relatively vast cultural forum on *Law & Order*. The need to fit more story elements into a standard television time frame has led to an accelerated narrative style that cannot accommodate expansive (and unrealistic) disclosures of unnecessary information about the characters' private lives. This focus on story over character has created an unexpected opportunity for the producers and writers to explore a range of cultural and political debates without having to defer always to the vision of any particular character; the characters are used as sounding boards for positions in these debates, but the point of the stories moves beyond the moral perspectives of the principal characters themselves. Finally, the lack of emphasis on any one character as providing a central point of view for the series opens up the door to changes in the cast. These cast changes further fuel and renew the opportunities for exploring conflict and contradiction within the individual episodes of the series.

Spin-Offs: The Forum and the Franchise

In addition to reaffirming the value of the story-driven narrative and helping reshape the domestic syndication market in the direction of one-hour dramas, Dick Wolf also used *Law & Order* to lead the way in creating prime-time brands that have a value beyond any one series. Wolf has done this by creating a number of spin-off series that have their own identities but are also directly linked to the *Law & Order* name: *Special Victims Unit* (1999–); *Criminal Intent* (2001–2011); and *Trial By Jury* (2004). Additionally, Wolf pushed the brand into newer media forms: the reality genre with a short-lived documentary series, *Law & Order: Crime and Punishment* (2002–2004) and CD-ROM computer games with *Law & Order: Dead on the Money* (2003), *Law & Order: Double or Nothing* (2004), and *Law & Order: Justice is Served* (2005). Finally, the brand has assumed global proportions as the original U.S. series are currently enjoying tremendous success internationally (as of 2010, the original series was sold in 220 territories; 240 for *Law & Order: Criminal Intent*; 190 for *Law & Order: Special Victims Unit*).[7] The success of these series internationally has led to the production of locally situated versions in the U.K. (*Law & Order: London*), France (*Criminal Intent*), and Russia (*Special Victims Unit* and *Criminal Intent*). The importance of these international series for the *Law & Order* brand demands a separate study that lies beyond the scope of the current project. In what follows, I will focus on the three primary U.S. based series that comprise the heart of the brand.

The first of these spin-off series, *Law & Order: Special Victims Unit*, was created in 1999 and deals with sex crimes.[8] The formula was originally intended to mirror that of *Law & Order*, but the centrality of the prosecutors' role was diminished in favor of the detectives, whose personal lives were also put on display more regularly (Dudsic 1999: A4). The second spin-off, *Criminal Intent* was an attempt to delve more deeply into the classical detective formula and create a character that is more like Sherlock Holmes than a typical police detective.[9] The prosecution's role is diminished even further in favor of delving into the mind and motives of the criminal, which are the primary concerns of the lead detective (Goren, played by Vincent D'Onofrio). Other major spin-offs have included *Trial By Jury* (2005–2006), which returned the prosecution to center stage. The narrative focus was on the efforts of the District Attorney's office to secure a conviction rather than the investigative aspects of the case.[10] This focus on the prosecution was attempted earlier as *Crime and Punishment* (2002), a reality series that followed the prosecution and the victims' families through the court proceedings of actual trials taking place in San Diego County. *Arrest and Trial* (2000) also offered a reality format, but with re-enactments. A final spin-off was *Law & Order: LA*, which mirrored the original *Law & Order* in its structure but transplanted the proceedings to California. In fact, most episodes begin after the suspect has been arrested and arraigned.

Given the variations on a theme that these different series represent, taken as a whole, the *Law & Order* franchise contains the potential for a rich forum on the criminal justice system in the U.S. While in the previous two sections I argue that the economics of the television industry – specifically the syndication market – helped shape and propel the cultural forum about crime that *Law & Order* represents, in this section I want to move in the opposite direction: to argue that, in other important respects, the potentials and limitations of a cultural forum about crime has driven the economic expansion of the franchise. I will discuss the development of different *types* of crime narratives within the franchise that have been developed in the U.S. alone in the form of spin-offs. The important questions have to do with how the *Law & Order* spin-offs work to address different kinds of questions about crime from within the franchise. To what extent does the franchise comprise a forum about crime? What are the means by which the various series are differentiated from one another? In order to address these questions, I want to turn to the issue of distinctions between the narrative foci of the three most successful series within the franchise. By looking at the varied narrative focus of each series, my analysis points to the potential for these different series to comprise a broader kind of cultural forum than is possible within any one series on its own.

This last point begins from Newcomb and Hirsch's insistence that the notion of the cultural forum encompasses "television as a whole system" that is larger than individual episodes, series, or even genres. What is at stake here is the "rhetorical slant" of different episodes, series, and genres as they come into contact with one another in the flow of the viewing experience rather than any one particular text taken in isolation (1983: 508–9). Clearly, Newcomb and Hirsch are working out an argument about the dialogic nature of television and are primarily concerned with the television *text*.[11] The key addition that I am making to their argument is that the television system is comprised of a dynamic set of economic and aesthetic practices that have implications for the idea of a cultural forum and how it operates.

Variations on a Theme

In his book *Adventure, Mystery, and Romance: Formula Stories as Art and Popular Culture*, John Cawelti makes a case for how different story formulas within the larger category of "crime literature" engage in a specific kind of cultural work. For Cawelti, formulas are "conventional ways of representing and relating certain images, symbols, themes, and myths" (1976: 20). More specifically, Cawelti argues that some formulas gain wide popularity and "become collective cultural products because they successfully articulate a pattern of fantasy that is at least acceptable to if not preferred by the cultural groups that enjoy them" (1976: 34). In short, these formulas engage with recognizable social concerns and cultural attitudes, which are then filtered through and articulated by a specific mode of representation; in the present case, the commercially supported one-hour television drama provides the largest representational frame. Within this larger frame, there exist several variations: police drama, detective drama, and legal drama, to name only the three most closely related to the present argument regarding *Law & Order*.

While Cawelti's sense of a *ritual* function of genres and formulas has largely been replaced by a more *discursive* and *institutional* approach to how genres function (as discussed in Chapter 3), his ideas about "patterns of fantasy" add a useful element to our own understanding of how a cultural forum about crime functions within the *Law & Order* franchise. What the *Law & Order* franchise offers is a complex array of sometimes complimentary and sometimes contradictory "fantasies" around the question of crime and justice in the U.S. This array of fantasies is constituted at several levels: the individual episode and competing ideas about a particular crime expressed therein; the individual series with varied approaches to justice between episodes; and the franchise as a whole. At the level of the franchise – which is the central focus of this section – each series has a particular narrative pattern and focus; the variations in these narrative patterns and foci provide the foundation (and at least the potential) for dialog between individual series.

Taking only the three most recognizable series in the franchise, we can see how these variations are structured as a larger cultural forum about crime. At the most basic level, we can use the titles of the series to demonstrate the key differences in focus among them. The flagship series, *Law & Order*, has the most abstract title and its narrative focus is similarly diffuse. In this series, with its split format, its decidedly story-driven narrative, and its reliance on the procedural, moral, and ethical complexities of the justice system, the "system" itself is at the center of the narrative: both the justice system and the larger social system within which it functions. Typically, while the murder victim remains vital to the case at hand (serving as the impetus for investigation), concerns about individual victims are often subordinated to the larger moral, ethical, and political questions of justice.

Two examples will help illustrate this point. In a third season episode titled "Sanctuary," a young black man is tried for murdering a white man during a race riot. While preparing to counter the defense's claim that the young man is actually a victim of a racist social structure, ADA Ben Stone asks his assistant, Claire Kincaide: "Would you drag an innocent man out of his car and bash his brains in?" Kincaide's response highlights the level at which the episode is operating, moving the question from the individual case in front of them to a larger systemic issue: "You know, I don't know how I'd react if I'd been screwed by the system my whole life." This movement from the individual to the systemic is made even more directly during a heated confrontation between Stone and defense attorney, Shambala Greene on the courthouse steps after Stone accuses her of trying to play on the jurors' sympathies:

GREENE: Sympathy is part and parcel of justice.

STONE: Right. And my sympathy is definitely with Mrs. DeSantis (the wife of the victim).

GREENE: Of course it is, because you can empathize with her. You have no idea what it is like through the eyes of Robert: to feel exploited; to be unemployed; to feel like you have no opportunities in this life. You pump enough air into a balloon eventually it goes "pop"!

STONE: And so the have-nots can take their frustrations out on the haves without recrimination? That's a hate crime!

GREENE: It's been happening the other way for centuries! Look, maybe you did march with Martin in the sixties, but you know what? Hanging a picture of Bobby on you wall just isn't gonna cut it anymore!

In another episode, "The Reaper's Helper" (season one), while trying a man with AIDS for assisting in the suicide of another AIDS victim, the defense attorney, Gordon, asks Stone in front of a group of reporters: "Mr. Stone, why do you want to put a dying man in prison?" Stone replies: "Mr. Gordon asks: do I want to put a dying man in jail. The answer is no. But we're asking a more important question in this courthouse: Does Jack Curry have the right, all by himself, to put a dying man in his grave?" Again, the question moves from the level of the personal (Jack Curry's health) to the systemic (in this case, the debate over the legality of assisted suicide in the case of AIDS victims). Throughout the course of the series, as the investigation of individual murders leads to debates about the nature of justice itself, the words of producer Rene Balcer ring true: "On *Law & Order*, the star is the system" ("All 3 Shows": 02E).

If *Law & Order* is about the system itself, the first spin-off, *Special Victims Unit*, is (as the title suggests) about the *victims* of crime. Whereas the victim of the crime in *Law & Order* is quite often reduced to merely the impetus for a larger investigation (a point made flesh by the consistently haphazard discovery of a dead body that begins each episode), the victims of *Special Victims Unit* are the whole point. And these victims include the police themselves as they cope with the "especially heinous" category of sex crimes.[12] As a way to differentiate the series from its predecessor, Wolf and his co-producers have introduced the personal lives of the detectives directly into this series. For instance, the first episode of the series ("Payback") follows the investigation into the murder of a cab driver who turns out to be a former Serbian officer who is under indictment for war crimes: in particular, the brutal rapes of dozens of women. His killers turn out to be two of these women and, in the course of the investigation, we learn that Det. Olivia Benson (Mariska Hargitay) is herself the product of a rape – an emotionally difficult fact that poses professional risks for her during rape investigations and almost gets her removed from the squad. In making the murder victim in this episode also a criminal in his own right, the series engages directly with the cultural dialog about who counts as a victim. This question cuts to the heart of what counts as mitigation, especially as it concerns what Wendy Kaminer calls "the abuse excuse" (1995: 11).

A later episode from the first season further highlights this centrality of the victim in a similar way. The episode title "Nocturne" focuses on the figure of a piano teacher who has been sexually abusing his young male students for years. In the process of trying to convince one of his older students to testify against the teacher, the police discover that this student has also abused a young child as well. This episode articulates the various sides of the "abuse excuse" and, importantly, never loses sight of the idea of the victim. Instead of using the victim as a means to get to bigger issues, the very idea of victimization is the issue. The victims in this case actually proliferate and assume center stage. To illustrate this point, the episode ends not with the indictment of the teacher (the original abuser) but with a demand by the second abused boy's father to indict the older student as well. The larger social debate about whether or not "abuse begets abuse" is not allowed to be the point of the episode. Instead, the focus remains on the victims of the abuse *despite* systemic complexities.

Finally, the third spin-off, *Criminal Intent*, places the relationship between the *criminal* and the detective at the center of the narrative. Unlike the other two series, which have "closed narratives" – meaning that we only see what the detectives see – each episode of *Criminal Intent* begins with the depiction of the crime itself. The viewer is aware of the perpetrator's identity from the beginning and has access to information that the detectives don't have first hand. Rather than exploring the nuances of the system, or how to define a victim, the focus of this series is the process of trapping the criminal by exploiting his or her psychological motives and/or weaknesses. More specifically the ability of the detectives, Goren in particular, to *classify* the criminal is central to this process.

An example from the first season illustrates this point. Having apprehended a member of a heist crew – the leader's girlfriend – Det. Goren tries to get her help in locating her boyfriend. He does this by analyzing and explaining their relationship:

> Your boyfriend was in prison a relatively short time but he couldn't control his sexual needs. You blamed yourself, didn't you? That you couldn't hold on to his loyalty and desire for 18 months. If I'm wrong in my evaluation, Miss De Luca, please tell me. He's betrayed you before, hasn't he? But he always came back, and you told yourself it was because you're the one that he loves. But he's never told you that he loves you, has he? Not in the way that would make you believe him. He makes demands on you: demands during love-making, demands that leave you unsatisfied. And you always submit, which for that reason he trusts you completely. You submit because you hope to earn his love and loyalty. Those hopes can only remain unfulfilled because Carl Atwood is a *narcissist*. He's incapable of loving you or anyone else. He seeks out partners to satisfy *his* needs.

This monologue (what producer, Rene Balcer calls the "aria")[13] occurs during the last fifteen minutes of the episode and works on two levels. First, it serves a narrative function of convincing the girlfriend to help them find Atwood. Second, it works to explain to the audience the deeper motivations for the crime in the first place, beyond the obvious impulses toward material gain.

In this regard, *Criminal Intent* functions more like a classical detective story than a police procedural. The classical detective formula contains two interlaced narratives: the story of the crime and the story of its solution. As Cawelti argues, "when the solution is announced, though technically the point of view does not change, in actuality we now see the action from the detective's perspective. As he explains the situation, what had seemed chaotic and confused is revealed as clear and logical." The classical detective story highlights the relationship between the reader and the writer that is forged through the narrative. "We are interested because we ourselves have been involved in the explanation and interpretation of the clues presented in the course of the investigation" (1976: 87–88). Rene Balcer, the executive producer of *Criminal Intent*, explains the series' relationship to the classical formula in this way:

> Goren is giving voice to whatever thoughts the perpetrator has, and he does this as a result of having spent the previous 35 minutes investigating

> this individual and learning what makes this guy tick. There's nothing that happens in those last eight or ten pages that hasn't already been teed up in the previous 45 pages. Goren doesn't pull things out of the air. That's the game of watching the show – wondering "Is this important or not?" The audience has the opportunity to see the "footprint," the clues, although Goren may not draw the same conclusions that the audience does.
>
> *(Littlefield 2005: 20)*

This sensibility of the relationship between the audience and the detective is in keeping with the idea that part of the pleasure of the classical formula is the "fascination" that "hovers about the detective's explanation as we measure our own perception and interpretation of the chain of events against his" (Cawelti 1976: 88).

Balcer's comments, however, may be slightly disingenuous considering the fact that Goren's interpretation of "what makes this guy tick" is always privileged (and confirmed) in the narrative. Balcer seems to suggest that there is an opportunity within the episodes themselves for disagreement between the audience and the detective as to issues of guilt, innocence, and justice. But the very structure of the episodes in the series, like the structure of the classical detective story, undercuts this kind of dialogic relationship by explaining the motivations of the criminal so fully – motivations that have been subtly alluded to in the course of the narrative.[14] Unlike the other series in the franchise, then, *Criminal Intent* offers a far more "closed" narrative in terms of explaining crime as "strictly a matter of individual motivations" which reaffirms "the validity of the existing social order" and ultimately turning "an increasingly serious moral and social problem into an entertaining pastime" (Cawelti 1976: 105). Having said this, however, I suggest that *Criminal Intent does* participate in the forum about crime by virtue of the fact that it offers a stricter vision of criminal motivations than the other series with which it comes into contact.

Of course, within and across the series, questions about criminals, victims, and the system itself necessarily overlap. Isolating them within each series has served merely to highlight the differences in narrative focus across the franchise. These differences in focus underscore the larger point that the franchise itself can and does constitute a complex forum about crime. In fact, the franchise has been actively *constructed* as a group of differentiated texts that, taken together, serve two functions: from a cultural perspective, they expand Wolf's ideal of treating complex issues from a variety of perspectives; from an economic perspective they expand his presence on the television schedule. The collapsing of these two functions is articulated by Wolf, who admits that "there's a part of me that would like to have a weeklong *Law & Order* marathon of the same story done by all three shows. Because they'd be completely different" (Levin 2002: 01E).

Conclusion

Law & Order has evolved from a struggling series into an expansive programming empire. The franchise that has grown up around it has been so vital to NBC's economic fortunes that some have speculated that the 2004 merger of the network and Vivendi-Universal was, at least in part, founded on the need for NBC to keep the *Law & Order* brand under its corporate umbrella (Carter 2003: C1). But the series and the franchise are more than shining illustrations of what success requires in the new television economy. *Law & Order* also demonstrates, perhaps more than any other set of texts, the extent to which television still constitutes a vibrant cultural forum about issues that are central to social life.[15] It is only by holding both economic and cultural concerns in balance that we can get a more complete understanding of the complex work that is *Law & Order*.

As I have argued in this chapter, *Law & Order* was initially conceived as a response to changing economic conditions in the U.S. television industry, but this response was primarily aesthetic: located in the narrative structure of the episodes and choices about how to develop characters over time. This episodic, story-driven approach to the first series, combined with an ensemble cast that could handle a high amount of turnover and a commitment to "current events" for story ideas, allowed the producers of the series to create a narrative world that could be easily consumed in small but full doses, each in the space of one hour, rather than demanding weeks, months, or even years of commitment from viewers. This model also allowed the writers and producers to pursue different angles on issues and place characters (regular recurring characters *and* episode-specific characters) in different combinations in order to extract the most dramatic conflict from each episode. The simultaneously stable and variable formula also served the series well in syndication, where less reliable viewing patterns dictate the need for flexible programming. This need for flexibility was nowhere more apparent than on emerging cable networks where *Law & Order* flourished in reruns, helping to solidify the economic viability of these cable networks, adding value (via increased ratings) to the episodes that were still being broadcast on NBC, and providing the primary impulse to expand the brand through spin-off series.

These spin-offs have continued to add value to the brand – each have earned solid ratings – but, more importantly for the purposes of this study, they have also expanded the possibility of a cultural forum about crime. This expansion is primarily a product of varied narrative foci between the series. By shifting the focus from the system (*Law & Order*) to the victim (*Special Victims Unit*) to the relationship between the criminal and the detective (*Criminal Intent*), the producers of these series created a forum within the franchise itself, as these series bump up against one another on the prime-time broadcast schedule, in syndication, and as characters cross over from series to series.

For Newcomb and Hirsch, the very idea of the cultural forum is founded on the idea of the "flow" of programming and the complex (and rather elusive) relationship between text and viewer. Rather than being an easily recognizable element of the television text, the forum is really a set of *possibilities* that are finally realized only at the level of viewer experience. This experience in the form of both programming practices and viewing habits has undergone an almost revolutionary transformation in the past twenty years. What *Law & Order* represents is a new set of possibilities – a new kind of cultural dynamics of television.

7

THIS COP'S FOR YOU

The Multiple Logics of the 21st Century Police Drama

As the previous case studies illustrate, the 1990s was an exceptional decade for the police drama on several levels. First, three supremely high-quality series dominated the genre: two of these (*Law & Order* and *NYPD Blue*) capturing significant audiences. Additionally, each of these series pushed at the boundaries of the genre in multiple ways. *Homicide* pushed at the uses and limits of realism, *NYPD Blue* pushed melodramatic form to offer a truly complex set of heroic figures and interpersonal dynamics, and *Law & Order* offered new narrative strategies for capturing the complexity and the multiple consequences of a single criminal act.

But the series were also *unexceptional* in that they illustrate the continued value of what police dramas have always offered. As I've been arguing throughout the preceding chapters, all police dramas engage, in one form or another, with the discourses of crime, community, and citizenship. Regardless of the perceived quality of the drama, these issues pervade stories about criminal behavior and the pursuit of justice. The differences are only partly a matter of kind: the particular issues that the writers and producers of a given series are willing or able to deal with. It is also a matter of the *degree* to which an individual series highlights the issues at hand. This issue of degree could also be stated as a matter of balance: the narrative balance between the issue and the drama and the degree to which they are intertwined.

Regardless of these differences between individual series, they each share a generic concern with tracing out the cultural contours of crime, community, and citizenship. Police dramas in the 21st century are no different. They approach these issues in thematic and structural relationship to the series that have come before, but from a slightly different institutional context. The emergence of premium channels like HBO and Showtime, as well as cable networks such as FX and TNT, as viable outlets for original dramatic programming have allowed writers and producers to explore cultural politics in ways that are not bound by the same mandates regarding content under which the broadcast networks must operate. Meanwhile, the police series on broadcast networks have largely embraced a mode of storytelling that might be called "high-concept" television: based on simplified and episodic storylines, distinct visual styles, and the potential for expanding franchises.[1] Of course, none of these elements is unique to the current moment. The three series discussed in the previous chapters each cultivated a distinct visual style, their episodes were comprised largely of stand-alone material with serialized elements strewn throughout and, in the case of *Law & Order*, had elements that could be expanded into future series. What is different in the new millennium is the degree to which very different approaches to stories about policing exist side by side, underscoring the inherent flexibility of the genre that has been the focus of this study.

This chapter focuses on three key series that together illustrate the variety in the police genre in the 21st century: *CSI: Crime Scene Investigation* (CBS), *The Shield* (FX), and *The Wire* (HBO). Each of these series respond to and reinterpret the police drama within specific institutional contexts ranging across network broadcasting, advertising-supported cable, and subscription-based cable. *CSI*, with its high-tech, high-concept look and its drive toward indisputable scientific truth, repudiates the moral (and stylistic) messiness of *Hill Street Blues* and *Homicide*. *The Shield*, on the other hand, builds on the messiness, offering a vision of justice overseen by an antihero who nearly obliterates the lines between criminals and the police, but also responds to (by repudiation) the forced moral clarity of *Nash Bridges*, the show for which Shawn Ryan, creator of *The Shield*, wrote for several seasons. Finally, *The Wire* responds to the perceived inability of police dramas on broadcast television to accurately portray the nuanced interplay between the "streets and the suites" by trying to deny any clear distinctions between legitimate and illegitimate action – long the narrative staple of police dramas – whether in terms of business, politics, or crime.

In what follows I consider a range of questions that address this larger issue of variety within the genre. The basic guiding questions of this chapter are: What are the different things that this genre can do, and under what conditions? In order to provide answers to these basic questions, I first address the shifting industrial locations of the police series in the 21st century. The focus here, though, is on the challenges to the "network logic" that defined the series of the 1980s and 1990s. In particular, I highlight the centrality of "branding" and the changing notions (both industrial and critical) of just what constitutes "television" in the first place. I then turn my focus to a comparative case study of the issue of "truth," especially as it animates the problem of "corruption," as constituted by the three series in question. By looking at and comparing the ways that each of the series handles the problem of locating "truth" (a necessary staple of the police procedural), we can see the variety (stylistic, narrative, and ideological) that informs the genre. If the previous chapters have demonstrated that this variety is nothing new to the genre, the current chapter underscores the idea that this variety is as much a product of institutional structure as it is the particular ideological concerns of the series creators. I want to be careful, however, about making too direct a connection between institutional location and ideological and/or artistic merit. Nothing about industry structure guarantees a particular kind of story or cultural politics. Instead, I see the changing conditions of the television industry not as mandating particular kinds of stories or series, but as *enabling* the variety that is the concern of this entire study.

What is Television? What is a Cop Show?: Network Logic, Branding, and Genre

In the introduction to *The Wire: Truth Be Told* (something of a series guide, but also a general reader about the series), David Simon tries to defend his series from the "cop show" label. "Swear to God, it was never a cop show. And though there were cops and gangsters aplenty, it was never entirely appropriate to classify it as a crime story, though the spine of every season was certain to be a police investigation in Baltimore" (Simon 2009: 1). But in an earlier letter to HBO on June 27, 2001, Simon suggests something slightly different about the identity of the series:

> It is grounded to the most basic network universe – the cop show – and yet, very shortly, it becomes clear to any viewer that something subversive is being done with that universe. Suddenly, the police bureaucracy is amoral, dysfunctional, and criminality, in the form of the drug culture, is just as suddenly a bureaucracy. Scene by scene, viewers find their carefully formed presumptions about cops and robbers undercut by alternative realities. Real police work endangers people who attempt it. Things that work in network cop shows fall flat in this alternative world.
>
> *(Simon 2009: 33)*

These two sentiments are not entirely contradictory; both point to a general dissatisfaction with the perceived limits of television police dramas – a subject that Simon had broached before with regard to *Homicide: Life on the Street*.[2] And both suggest that these limitations were an outgrowth of the logic of broadcast networks. Later in his letter to HBO, Simon argues that it would be "a profound victory for HBO to take the essence of network fare and smartly turn it on its head, so that no one who sees HBO's take on the culture of crime and crime fighting can watch anything like *CSI*, or *NYPD Blue*, or *Law & Order* again without knowing that every punch was pulled on those shows" (Simon 2009: 34).

What the previous chapters have argued is that the commercial and regulatory logic of broadcast networks have not limited the scope of the cultural forum on crime, community, and citizenship, but that they do have a necessary effect on the style in which those elements are confronted. The new logic of which Simon speaks is not necessarily a challenge to the "police drama" as a category, but to the network logic that shapes the forum in particular ways, under historically specific conditions. Those conditions are changing as new delivery technologies and expanded distribution outlets complicate the regulatory and commercial bases of the network logic. The new logic allows for a more expansive, comprehensive, and challenging type of storytelling, but it does not challenge the baseline of what police dramas are about. The clear distinctions that Simon draws are, on the one hand, a useful rhetorical device for selling HBO on his project. On the other hand, they raise the question of just what a cop show is in the first place. Simon's claims highlight once again the idea that police dramas are inherently conservative ideological set pieces. Logically, then, if a series challenges these politics, it must not be a police drama, or to be such a subversive version of that world as to render it mostly outside the genre. One of the goals of this project has been to move beyond distinctions based on taste, perceived quality, or ideological perspectives to argue that all police dramas trade in the kinds of concerns that animate a series like *The Wire*. The differences between series are matters of degree rather than kind.

Importantly, the series of the 21st century thus far are products of differing institutional arrangements that open up (or deny) an even wider range of aesthetic, thematic, and political possibilities. While *Hill Street Blues* and *NYPD Blue* in particular pushed at the boundaries of what was acceptable to say and show on network broadcast television, neither series was able to approach the grittiness of *The Shield* which, as a product of a cable network (FX), was largely free from FCC oversight regarding indecency, beholden primarily to the concerns of advertisers. And none of these series benefitted from the free rein enjoyed by *The Wire*. As a premium pay channel available only via cable and satellite, HBO answers primarily to subscribers and experiences no federal or local regulation in terms of content. Thus, *The Wire* (like *Oz* and *The Sopranos* before it) was able to explore its world of drug-trade bureaucracy and government corruption unblinkingly. Both *The Shield* and *The Wire* were designed for smaller, "quality" audiences, while *CSI*, a broadcast network offering, still must attempt to capture as large (though not entirely undifferentiated) an audience as possible and keep an eye open toward future syndication possibilities (a possibility only partially open to *The Shield* and *The Wire*).[3] The narrative structure of *CSI*, like *Law & Order* before it, bears this distinction out most clearly: it is highly episodic and therefore can be consumed as stand-alone stories and later stripped in syndication without regard for running order.

What this interplay between the stylistic and thematic elements of these three series underscore once again is the variety that defines the police drama. The key differences between the periods 1980–2000 and 2000–2010 are the institutional "locations" of these various programs. In the earlier period, the series were all produced for broadcast networks and operated loosely under the same network logic. The products were differentiated, to be sure, but the guiding pressures were largely the same. But with the emergence of cable networks as viable sites for popular (and award winning) original series programming, and with pay channels looking to develop long-term brand identities through series programming, the aesthetic and thematic range of the television police drama has begun to change again. If *Hill Street Blues* demonstrated what the broadcast networks and production companies were capable of achieving by imagining their audience differently, *The Shield* and *The Wire* pushed this new logic to its extreme. At the same time, however, the old logic has not given way completely. The undeniable success of *CSI* and its franchise partners (*CSI: Miami* and *CSI: New York*) demonstrates that the broadcast networks are still formidable players in the media landscape. What has changed for the broadcast networks is that they are now one part of a larger, highly differentiated media production circuit, connected to the cable networks and pay-cable outlets via their parent companies, but operating on different parts of the potential audience.

These changes highlight a central problem for critics interested in understanding the cultural work of series, genres, and networks: that of defining television itself. As Charlotte Brunsdon has pointed out, the "television" that is the object of television studies is "a production of the complex interplay of different histories – disciplinary, national, economic, technological, legislative – which not only did not exist until recently, but is currently, contestedly, being produced even as, simultaneously, the nationally regulated terrestrial broadcasting systems which are its primary referent move into convulsion" (1998a: 95). What Brunsdon suggests here is that the object of "television" itself cannot be understood outside of a range of competing contexts for understanding what it is in the first place. A scholar interested in the regulation of television will have a very different understanding of what "television" is than someone whose primary concerns are the ideological work of television texts or the resistant work that audiences can do with those texts. Furthermore, the very structures that produce the texts we know as "television" are in a constant state of flux and redefinition, forcing scholarship to continually renew our understanding of what "television" we are talking about.

Picking up on Brunsdon's arguments, Amanda Lotz points out that the "development of cable as a broadcast competitor differentiated by a variant economic model and regulatory status is central to the disruption of assumed understandings of television" (2007a: 114). For Brunsdon and Lotz, matters of commercial versus public or state-run systems, as well as major shifts within the economic, technological, and regulatory underpinnings of commercial television systems alone, force us to reconsider our assumptions about how "television" operates as a set of cultural practices. These changes should also encourage us to reconsider the cultural work of genres such as the police series. The changing logics of television itself necessarily alter the cultural, aesthetic, and political logics that animate genres as well and create distinctions that go well beyond evaluations of worth or quality.

The first area within which we might note clear distinctions within television – and, by extension, the content that audiences consume – is in the area of *content regulation*. Under the rules first enacted in the Radio Act of 1927, the FCC (at the time the FRC, which changed to the FCC in 1934) has the power to regulate the content of broadcast radio and television in order to ensure that broadcasters act in the "public interest, convenience, and necessity."[4] This brief phrase covers a great deal of regulatory ground regarding broadcaster responsibility and what it means exactly has been widely debated in the legislatures, the courts, and among media scholars. Based as it is (at least partially) on the fact of the scarce and valuable physical commodity that is the electromagnetic spectrum, the phrase is perhaps intentionally vague in order to allow for flexibility in the face of technological change (not to mention shifts in the reigning political ideology). Put simply, broadcasters are granted licenses to use and profit from this spectrum and, in return, are required to uphold "community standards" by avoiding obscene or indecent materials. These rules have had a lasting impact on the production of television fiction, establishing a relatively narrow range of thematic, narrative, and aesthetic choices available to producers.

At the same time, cable networks, both commercially supported (e.g. TNT, FX) and subscription-based (e.g. HBO, Showtime) exist largely outside the purview of the FCC with regard to content. While obscene material is forbidden regardless of the outlet, the area of indecency (poorly defined as it may be) is handled differently for cable and broadcasters. While broadcasters may be fined by the FCC for airing any content deemed indecent during the "safe harbor" hours of 6 a.m. to 10 p.m., cable networks operate under the discretion of the cable service provider, which is granted the right by the FCC to determine which cable networks (i.e. programming content) it wishes to make available to viewers. The FCC still retains certain areas of authority over cable content, such as a ban on tobacco advertisements and the requirement that all subscription-based channels not paid for by a customer must be fully scrambled, but restrictions about language, violence, and sex are left largely up to the cable system providers.[5] The result is that programming originating on either commercially supported cable networks like FX (*The Shield*) or subscription-based networks like HBO (*The Wire*) is granted far more leeway in terms of manifest content than programming produced for broadcast networks (*CSI*).

Of course, these distinctions are beginning to break down in the face of the expanded competition that cable itself represents. New and emerging distribution technologies and increased competition among a wider range of competitors for audience share – such as YouTube, Netflix, and Amazon Instant Video – have encouraged broadcasters to push the envelope on content. Like cable and satellite, these new outlets for video content are not bound by the same FCC rules regarding indecency and have created pressure points for broadcasters trying to remain competitive for younger audiences that might be seeking more daring content. Additionally, the ease with which sites like Facebook and any number of blogs can circulate this content widely places further pressure on broadcasters and, more importantly, erodes the distinctions viewers make between broadcast, cable, and internet-derived content.

A second issue for understanding what television is involves *access* and the experiences of different types of viewers. In the network era that lasted roughly from the beginning of network radio to the middle of the 1980s (when the three networks saw a turnover in leadership for the first time), access to broadcasting was largely unified in the sense that there were limited distribution outlets and a relatively stable set of technologies for viewership.[6] In the post-network era, however, the variety of viewing experiences is part and parcel of the economic structure of the industry. As Barbara Selznick points out, audiences (now imagined in micro-units as opposed to large, undifferentiated blocs) are valued for "their willingness to follow texts across media platforms," pulling media content into different devices: their televisions, their laptop computers, their smart-phones (2009: 178).

But old ways do not die so easily. Selznick argues that the conditions that define the post-network era in television place dual demands on television networks and the conglomerates that own them:

> At the beginning of the twenty-first century, the television industry is caught between an unpredictable future and a waning past. Executives may recognise the need to target audiences with texts that will encourage them to "pull" television shows from numerous distribution windows, but they also realise that they still need to grow viewership in order to maintain advertising rates for their "push" television networks.[7]
>
> *(2009: 178)*

The situation that Selznick describes might be explained in two related ways. First, as Michael Curtin has shown, there are two tendencies that operate within the contemporary culture industries. One tendency focuses on content aimed at broad national or global audiences and requires relatively low involvement; the other seeks out niche audiences that are "more likely to be highly invested in a particular form of cultural expression." In this scenario, the bifurcation of industry strategies is a product of "flexible corporate frameworks" that "connect mass market operations with more localized initiatives" (Curtin 1996: 197). A second way to explain the situation that Selzick describes is what Christopher Anderson calls the "burden of history." As Anderson notes, "the constitutive choices made during an earlier period of American television inhibit change at the broadcast networks." Anderson is referring to the fact that broadcast networks in particular have remained remarkably committed to a certain style of programming, one constructed long ago and built around culturally and historically specific definitions of what should constitute the broadcast schedule, the broadcast season, and the structure of series narrative (2005: 78). At the same time, thanks to the steady incursions of alternative programming outlets, the broadcast networks know that they can no longer rely on a mass audience approaching anything near the numbers they once commanded. These networks move toward innovation, but are held in check by the weight of programming practices that have come to seem like the natural order of things.

For each of these scholars, the direction that the culture industries are taking (television in this particular case) is toward the concept of branding, whether at the level of a series franchise or the network itself (or both). Branding, of course, is nothing new to commercial television. From the beginning, the U.S. networks created identities for themselves through their programming – in the types of performers and series they developed and fostered and, perhaps especially, through their news operations. Similarly, like the major Hollywood studios before them, independent production companies have typically cultivated themselves as brands for certain kinds of programming. From Jack Webb's Mark VII in the 1950s and 1960s, to MTM and Norman Lear's Tandem Productions in the 1970s and 1980s, and Aaron Spelling in the 1990s, these companies turned out a significant amount of series programming for the networks with a remarkable consistency across products in terms of structure, style, and subject matter. The series franchises that have come to dominate primetime drama on the broadcast networks in the post-network era (*Law & Order*, *CSI*, *NCIS*) emerge from the same tendencies toward creating a recognizable brand through a closely related range of products, but they differ from the earlier era in that the various series are tied together under a single franchise label that, in turn, exists under the larger brand of the production company (e.g. Wolf Films, Jerry Bruckheimer Productions, Belasarius Productions).

In the era of cable and satellite – an era of rapidly expanding viewing choices – the concept of branding at the level of the network has become even more crucial. As Greg Moyer, chief executive and creative officer of Discovery Communications which operates seventeen different cable networks, pointed out: "We no longer think of ourselves solely as distributors, but that the heart and soul of our business is our brand" (Littleton 1996). However, cable networks and broadcast networks think about building and maintaining their brands in slightly different ways. For cable networks, there exists a necessary tension between off-network "stripping" of syndicated series and creating their own original programming. Networks like TBS ("Very Funny") and TNT ("We Know Drama") originally built their brands by focusing on a specific type of syndicated programming: situation comedies and one-hour dramas, respectively. For cable networks with relatively low visibility, tried and true syndicated programming is a natural way to build an audience. At the same time, original programming can build a higher profile and a stronger, more coherent brand for a struggling network than a hit drama that has been associated with a number of different networks (Sherman 2003).

Of course these two strands of branding are deeply intertwined. Most cable networks build original series and syndication strategies to complement one another (Albiniak 2008). Networks use syndicated series such as *Law & Order* to build audiences that might then find their original programming. For example, the success of TNT's original series, *The Closer*, relied significantly on the establishment of an audience for one-hour crime dramas through the programming of *Law & Order* and *NYPD Blue* reruns under the network's slogan, "We Know Drama." (Albiniak 2008). Alternatively, USA network has established its "Characters Welcome" brand through the creation of "blue sky" original series like *Monk*, *Burn Notice*, and *Psych* as well as the repurposing of syndicated series such as *House* and *Law & Order: SVU*. According to USA's Vice-President for Marketing, Chris McCumber, for any series on USA, "having a central character with a unique skill set is key, as is being very blue-sky and positive. Where a lot of other places are going dark and crazy-edgy, we're much more blue-sky and aspirational" (Hampp 2010). Even acquired series that have a darker edge can be made to fit the USA brand. As McCumber explains:

> Even with the acquisitions we've made like *House*, Dr. House is a quintessential Characters Welcome personality. And when we created the Characters Welcome brand we already had *Law & Order: SVU* within the family and *SVU* is a fairly dark, tough show. So the way we marketed it was focused on Mariska Hargitay's character and telling those stories through her eyes and attaching a little more to her backstory. We gave it a new spin we hadn't seen before.
>
> *(Hampp 2010)*

What these differing branding strategies point to is the fact that in the post-network era of television, networks can no longer rely on programming for a mass audience. Programming, whether original or syndicated, must now be carefully aligned (even retrofitted) to the demographic segment of the audience that the networks and advertisers wish to attract most.[8]

How then do the three series that are the focus of this chapter represent this attention to branding? And to what end? Each series is identified with a network that has a distinct brand identity. In each case the series has either been instrumental in shaping that identity or – in the case of *The Wire* – maintaining the brand momentum established by a previous series (i.e. *The Sopranos*). Finally, the series taken together demonstrate the variety within the genre as it stretches out across different network brands that play to different demographic segments of the audience and are produced, distributed, and consumed within variable institutional frameworks: a situation that I argue actually extends the potential for the "cultural forum" on crime, community, and citizenship. As Amanda Lotz has argued, new and emerging institutional structures and practices need not diminish the validity of thinking about a "cultural forum," especially if we acknowledge the changed scope of "television" as a medium:

> The cultural forum model remains of great value if we define television as it now exists as a diffuse medium, rather than narrowly confined to an

> individual series, episode or similar unit. Understanding that we can only speak of an aspect of television – not television as a whole – and that we must comprehend the breadth of television, yet speak of it with specificity, helps reintroduce the pervasiveness of its messages that now reach a more narrow scope at the level of the individual programme.
>
> *(2004: 431)*

In the case of *The Shield*, *The Wire*, and *CSI*, the variety that these series represent connects the relative narrowness of brand identity with the increased scope of the police drama across an expanded television landscape. In what follows, I first illustrate the connections between these series and their respective network brand identities. I then provide an analysis of how each series constructs a relationship between police work and "truth" and its concomitant relationship to the problem of corruption. My point is to illustrate the breadth and depth that the genre offers with regard to a particular issue in order to demonstrate the relatively large scope of the cultural forum. The diffusion of the viewing options (emphasized by the centrality of variously narrow network brand identities) does not diminish the potential for a "cultural forum" around crime; instead, it reconstitutes the forum at the level of genres and series that are attached in multiple ways to narrow and shifting brand identities.

FX and The Shield

In the case of FX's *The Shield*, the series was a significant part of the network's rebranding strategy. FX began network life as fX and branded itself as "TV Made Fresh Daily," offering a slate of live talk shows and relatively tame reruns from the 1970s and 1980s (e.g. *Eight Is Enough* and *Hart to Hart*). This iteration was largely ignored and the network was re-launched beginning in 1998 through the acquisition of edgier in-house Fox series, such as *NYPD Blue* and *The X-Files* as well as original talk, variety, and game shows aimed at a specifically male demographic (Moore 1999). This move toward a primarily male demographic was then extended into original one-hour dramas, the first of which was *The Shield*. What *The Shield* represented was the network's strong move away from its reliance on traditional broadcast network fare, its unabashed appeal to a male audience, and its willingness to push at the boundaries of acceptable content.

The Shield premiered on 12 March 2002 amidst an array of protests by watch-dog groups like the Parent's Television Council (PTC) and threatened boycotts by advertisers. In April 2002, *Broadcasting and Cable* quoted PTC founder, L. Brent Bozell as saying that the "PTC has launched a targeted campaign to expose the companies that continue to support the unprecedented filth FX has chosen to display on *The Shield*" ("PTC Attacks '*Shield*'"). High-profile advertisers such as Burger King, Office Depot, Anheuser-Busch, and New Balance dropped out after the series premier, following General Motors and John Deere, which had dropped the show prior to its premier. Perhaps because of this public controversy, the series premier enjoyed an exceptional 4.1 rating, but slipped to 2.2 by it's fourth episode (Dempsy 2002). While this slippage might have caused some concern among both advertisers and executives, 65 percent of those viewers were adults aged 18–49 and primarily male: a demographic that advertisers covet. Despite the declining numbers, but because of the good demographic profile of the series, FX was selling 75 percent of its advertising slots even with the defections. As Peter Liguori, President of FX at that time stated, "We're selling the show out because it's a value for a very, very specific and very tough audience, and that's an audience that seeks discerning television. This show is a show for adults" (Owen 2003). Clearly, the series had successfully captured the audience that the network was seeking in its rebranding efforts.

Interestingly, the series was not initially what the network was seeking for its first foray into original dramas, largely because of its generic identity. Shawn Ryan, creator of the series, noted that the only rule the network gave him in creating a series was: no cop shows. As a writer for *Nash Bridges* (CBS), Ryan was well acquainted with the perceived shortcomings of the form: sensationalistic and exploitative scenarios, cops who were too virtuous, criminals who were too villainous. "*Nash Bridges* was sort of a joyride through San Francisco more than it was a serious cop show. I would see police officers handling horrible traffic accidents, and they were very solicitous to the victims, but then off on the side they'd be making jokes among themselves because, well, they'd seen it before. But I couldn't use that stuff" (Garvin 2002). *The Shield* was an attempt to represent the work of policing as contradictory: heroic work compromised by politics and individual failings.

By moving away from some of the more traditional concerns of the form, *The Shield* presented itself as the appropriate vehicle for a niche network looking to distinguish itself in the crowded media marketplace and opened up the floodgates for cable networks looking to extend their brands through original programming. Current president of FX, John Landgraf points to the significance of *The Shield* for television as a whole, not just for FX: "It was a very audacious bet. At that point in cable history, only HBO had had any success with scripted series. And basic cable had never before attempted the type of high-quality serialized dramas for which HBO was renowned" (Pierce 2008). In the wake of *The Shield*'s success, the network has built a roster of series aimed at its young, male demographic (*Rescue Me*, *Justified*, *It's Always Sunny in Philadelphia*, and *Sons of Anarchy*), that bear more resemblance to each other as "FX series" than they do to other examples of their various genres. Similarly, other advertiser supported cable networks have followed suit. Most notably, TNT, USA, and AMC have produced and programmed original series (such as *The Closer*, *Burn Notice*, and

Mad Men, respectively) that continue to redefine the contours of commercial television and build their particular brands.

HBO and The Wire

These commercial cable networks came to original series programming behind the lead of HBO, which had already established itself as a force for original dramas that challenged the traditional notion of genre with series such as *Oz* and *The Sopranos*.[9] With the premier of *The Wire*, HBO was not attempting to remake itself as much as build upon an already established model of success. That model is predicated on the network's position as a subscriber-based channel outside of commercial broadcasting and cable. David Simon's letter to the network, described briefly at the beginning of the chapter, plays into that model in specific ways, and frames *The Wire* as a series uniquely suited to HBO's brand: "It's Not TV. It's HBO."

The key to this brand identity is, of course, exceptionalism. This issue of exceptionalism is a matter of specific ideas about "quality," particularly as they attach to notions of auteurist sensibilities (i.e. an avowed celebration of television writing), and the ability of a series to be "groundbreaking," especially the degree to which the content of a given series would be *in*appropriate for broadcast television. As Deborah Jaramillo points out, the development of early HBO series such as *The Larry Sanders Show*, *Sex and the City*, *Oz*, and *The Sopranos* involved "the systematic co-optation of traditional broadcast formats (hour long dramas, half-hour long sitcoms) through the addition of a pay-cable sensibility – more leeway in the area of explicit content and no commercial interruptions" (2002: 63). The inherent advantages of freeing up writers and producers with regard to more explicit content lends itself well to an auteurist and exceptionalist celebration of unbound creativity, even as the forms themselves adhere fairly closely to the norms established by traditional broadcast formats.

Adhering so closely to traditional broadcast forms, such as the police drama, allows a series like *The Wire* to tout what Jaramillo calls "generic prestige" (2002: 67). For Jaramillo, writing about *The Sopranos*, the series was unconventional (and therefore prestigious) in at least two ways. First, prior to *The Sopranos*, there were virtually no significant gangster dramas in the history of U.S. broadcast television. Thus, *The Sopranos* stood out as a distinct television form, though as Jaramillo is careful to point out, the series is actually more closely tied to traditional television forms of the "workplace family" drama and family melodramas. Because of this formal distinction, the series also traded on intertextual references not to other television series, but to the more culturally prestigious gangster film (e.g. *The Godfather*, *Public Enemy*, *Goodfellas*), creating the impression of a more prestigious cinematic pedigree. This adherence to a sensibility of "prestige" is, of course, central to the HBO brand.

In a similar fashion, David Simon and HBO worked hard to set *The Wire* apart from the abundant and mostly dismissed television police drama:

> The first thing we had to do was teach folks to watch television in a different way, to slow themselves down and pay attention, to immerse themselves in a way that the medium had long ago ceased to demand...And we had to do this, problematically enough, using a genre and its tropes that for decades have been accepted as basic, obvious storytelling terrain.
>
> (*Simon 2009: 3*)

This "different way" of watching television is, of course, more in keeping with the somewhat questionable trope of how people supposedly watch feature films (with rapt attention and an appreciation for the finer points of character and story) than for the injurious commercial interruptions that are the lifeblood of advertising-supported television. But another part of this equation for Simon was to push HBO itself in a new direction. In his letter to Chris Albrecht and Carolyn Strauss at HBO, Simon laid out what he saw as the value of a series like *The Wire* for the network. According to Simon, having brought "fresh worlds" (prisons, gangsters, and drug corners) to television, HBO was now in danger of falling into formula. *The Wire*, according to Simon, presented a new opportunity for HBO to extend its brand and compete directly with the broadcast networks. For Simon, *The Wire* represented a challenge to broadcast "network logic" and "the next challenge for HBO" (2009: 33). The generic prestige that Simon was after was not the same as *The Sopranos*, which was about creating a "fresh world" on television. Instead, *The Wire* was about challenging the basic assumptions of a tried and true network television form within the more open (and more prestigious) confines of the HBO brand.

CBS and CSI

Both *The Shield* and *The Wire* represented a form of generic distinction and a way of pushing their respective network brands in new directions; *CSI* represents much same impulse but in quite different ways. Most obviously, *CSI* uses visual style as a marker of distinction. As Tom Steward has pointed out, *CSI* is both a highly standardized product and one that stands out from the rest of the television landscape. The series, along with its franchise brethren, is standardized in terms of its basic plot structures and narrative strategies. At the same time, the series is constructed around a color palette and editing style that sets it apart from much of prime-time programming. Describing the look as emanating from producer Jerry Bruckheimer's sensibility, series creator, Anthony Zuiker points out that "(the show's) colors are richer, the palette, the lighting, dark and edgy, and the speed is quicker" (Booth 2005). Additionally, the series makes significant use of CGI technology to create what has come to be known as the

"CSI shot": a visual recreation of forensic evidence (e.g. the path of a bullet or other foreign object) from within the body of a victim. The impulse behind the look of *CSI* was mostly a response to a more crowded marketplace for programming and expanded viewing choices. Creating a style that exceeds that of other dramas was necessary to "capture the floating viewer traversing television flow" (Steward 2010: 735). As Derek Kompare argues, *CSI*'s heightened visual style:

> is appropriate to twenty-first century media culture, and particularly to dramatic television. In a fast-paced, screen-centered media age when cutting-edge CGI regularly illustrates news and documentary programming, and nearly 2 billion people routinely use the multiple-layered systems of representation on mobile phones, web browsers, and other platforms, mediated information normatively circulates in many ways.
>
> *(2010: 425–426)*

While almost all police dramas, from *Dragnet* onward, have relied on visual styles that work to heighten the sense of verisimilitude, *CSI* takes the next logical step and attempts to construct ultra-realistic representations of evidence (including the internal working of the human body) while simultaneously crafting a color palette and editing style complete with CGI effects. This seems to define what John Caldwell refers to as "televisual excess" which, Caldwell argues, often serves to legitimate the authority of the televisual text – in this case, the authority of a particular vision of crime and criminal justice more in keeping with the kinds of screen-based imagery that simulated an increasing number of viewers' tastes and experiences (1995: 190).

Another way that *CSI* works to enhance the brand identities of its networks (both its original broadcast airings on CBS and its syndication homes – currently USA, Spike, and TV Land in the U.S.) is through its highly formulaic series structure. Like *Law & Order* before it, the formulaic structure of *CSI* and the other series produced by Jerry Bruckheimer – notably, the rest of the *CSI* franchise, as well as *Without A Trace* (2002–2009), and *Cold Case* (2003–2010) – works in two ways. First, because of it episodic structure, it can be syndicated easily across a range of networks looking to build their brand with a quality series. In the case of *CSI*, the series has served several cable networks in syndication, demonstrating that series identities can be flexible with regard to fitting into a network brand identity. For instance, the USA network website highlights the specific characters featured in the series, including significant guest star appearances (to fit with their "Characters Welcome" brand) while Spike, a network aimed at a primarily young, male demographic, created "Jerry Bruckheimer-style" promotional spots for the series that emphasized "action, gross-out elements, and the flashiness of its Sin City setting" (Karrfalt 2005). The ratings for *CSI* on CBS, especially among males, increased significantly after it was stripped on Spike, and this growing success of *CSI* on CBS encouraged the network to build a schedule that included more episodic crime series with similar stylistic and structural approaches.[10] As Derek Kompare has pointed out, in "the 2008–9 season, CBS had no less than 11 regularly scheduled crime dramas on the air, half of its total prime-time schedule" (*2010:* 1515–1516). Several of these series were produced by Jerry Bruckheimer as well, and provided character crossover opportunities between shows, "transforming much of CBS' schedule into a coherent "universe" of characters and settings, all emanating from the original *CSI*" (Kompare 2010*:* 1522–1523). Thus, the formulaic/episodic structure of the series allows for a certain flexibility in syndication, which can then lead to greater ratings success for the original run of the series on its home broadcast network, encouraging in turn a stronger commitment to similarly styled programming – the building blocks of a successful brand.

The element of coherence points to the second way in which the series' formulaic structure works to enhance brand identities: the establishment of a franchise built around either a series brand (like *CSI*) or an authorial figure such as Jerry Bruckheimer. As stated earlier, Bruckheimer's presence on the CBS schedule has been significant, with five dramatic crime series (all three *CSI* series, as well as *Without A Trace* and *Cold Case*) on the schedule at various points since 2000. In much the same way that CBS created an internally coherent programming schedule around a large number of similarly styled crime series, Jerry Bruckheimer's production company, Jerry Bruckheimer Television, uses standardization of product to tie together a "potentially disparate group of texts with a product label or guarantee." While Bruckheimer's strategy for these crime dramas is useful for a producer looking to "cultivate internal consistency for brand differentiation in the market" (Steward 2010: 740) and move product across a range of outlets, CBS has also capitalized on this consistency to build its brand. One of the ways that CBS has been able to create this consistency is by co-owning the *CSI* franchise.[11] The ability of CBS to have an ownership stake in the property is a direct outgrowth of the demise of the Financial Interest and Syndication Rule (also known as fin-syn), which had limited the degree to which broadcast networks could own (and thus be able to sell in syndication) the series that they programmed. Fin-syn was abolished in 1993 in light of the growth of competing broadcast networks (i.e. Fox) and the continued growth of cable outlets. While fin-syn allowed independent production companies to establish themselves as quality brands within a relatively narrow marketplace, the demise of the rule in the wake of the expanded marketplace of the post-network era, encouraged and enabled the major networks to more carefully and strategically brand themselves – a prospect that could be accomplished most easily through programming that they owned and could control more completely. Within this context, formulaic series like *CSI* that could be expanded through a particular brand identity and then syndicated through multiple partnerships across the broadcast and cable landscape became a centerpiece of network programming strategies and brand construction in the post-network era.

The Truth Economy and Corruption in *The Shield, The Wire,* and *CSI*

A particular concern with police dramas is the question of truth and the ability to explain crime in ways that make sense to viewers. But, taken as a whole, police dramas trade in notions of the truth that are as varied as the series themselves. In particular, the notion of certainty and how it might be arrived at is a structuring problem in most police dramas, whether episodic procedurals or serial melodramas. The cultural life of Jack Webb's series *Dragnet* has been reduced in popular imagination to a few stock sounds, phrases, and images: the theme music, the staccato dialogue, the spare sets and close-up framing, and perhaps more than anything, Joe Friday's oft-repeated admonition to witnesses instructing them to stick to "just the facts." On the opposite side of the divide, one might consider the opening sequence of the pilot episode of NBC's *Homicide: Life on the Streets* ("Gone For Goode") in which the very notion of searching for truth or explanations is suggested to be potentially pointless:

LEWIS: If I could just find this damn thing I could go home.
CROSETTI: Life is a mystery. Just accept it.
LEWIS: You're in your own world, Crosetti.
CROSETTI: It's the questions…not finding, looking. I read about it in this book.
LEWIS: Now, since when did you ever read a book?
CROSETTI: I read this book….an excerpt of this book.
LEWIS: See that's what I'm saying. You said you read a book, but you didn't read nothin' but an excerpt.
CROSETTI: It says that you never really find what you're looking for because the whole point is looking for it. So, if you find it, it defeats its own purpose.
LEWIS: You know you're in your own little world because don't nobody want to live there with you.
CROSETTI: You try to explain everything, you know? There are things you cannot explain.

For Lewis, the truth (the thing he is looking for) is a means to a clear end. For Crosetti, on the other hand, truth or certainty is an unattainable endpoint – a matter of negotiating with facts in ways that never illuminate a final solution but raise only more questions about larger, more systemic issues. I argue that these two positions (presented here as humorous and equally untenable) represent the range within which the police drama works with regard to the question of truth. Additionally, this more philosophical issue of truth attaches to the very real problem of corruption, which since at least the 1960s has been an ongoing, if often unstated concern of television police dramas. In what follows, I offer brief illustrations of the connection between truth and corruption as played out in *The Shield*, *The Wire*, and *CSI* in order to illustrate the kind of ideological variety that I have been concerned with throughout this project.

Multiple Layers of "Truth"

With regard to the issue of *truth*, each of these series works on at least three levels. The first of these levels is the series' adherence to a realist aesthetic – a standard trope of the genre. As with *Hill Street Blues*, *Homicide*, *Law & Order*, *NYPD Blue*, and so many other series before them, *The Shield*, *The Wire*, and *CSI* each constructs a representational style that highlights the "reality" of what viewers can see and hear. In the case of *The Shield* and *The Wire*, the series' aesthetics highlight messiness (handheld cameras, location shooting, lots of natural lighting) and grittiness (overtly violent action, harsh language, sexuality, drug use) in the service of creating a more "immediate" sensibility. But the grittiness of these series is, again, a difference of degree rather than kind and largely enabled by their institutional locations on FX and HBO, respectively. The look of *CSI*, on the other hand, is largely predicated on a tension between the unreality of both the "glitz" of its Las Vegas setting and the clearly computer generated graphics of the "CSI shots" on the one hand, and the presentation of forensic fact on the other. As Deborah Jermyn points out:

> Fostering a potent kind of intimacy between the television audience and the body, CSI captialises on the small screen space and the intimacy of the medium by actually taking the viewer into the body's "inner space."...But at the same time, these sequences and their deployment of CGI – labeled the "*CSI*-shot" by the programme's producers – are clearly about *spectacle* and capture more of the physical drama of the body's interior than any real medical probe ever could.
>
> *(2007: 80)*

In fact, each of the series in question negotiate this tension between reality and spectacle, though perhaps not as overtly as *CSI*; the "realism" of *The Shield* and *The Wire* is as much a construction as the CGI of *CSI*. But through these careful aesthetic constructions, each series makes a claim to a kind of truth about what police work, criminal activity, and evidence look and sound like.

A second level of "truth" on these series is narrative and involves the general facts of the cases at hand. This may seem rather obvious on the surface since all police dramas are concerned, either centrally or peripherally, with investigations that reveal facts and solutions based on those facts. But each series treats facts – and the truths they represent – in slightly different ways. In each series, the discovery of facts constitutes the arc of each investigation. In the case of *The Shield* and *The Wire*, smaller investigations can overlap with and inform larger story arcs. For example, in the pilot episode of *The Shield*, a suspected child molester refuses to admit to his acts, even when presented with specific facts that link him to the victim. The detectives doing the interrogation then turn the suspect over to a third detective, Vic Mackey, who beats the confession out of the suspect. On the one hand, this single case presents us with facts that lead to a truth (the suspect is clearly guilty). At the same time, the methods to which the detectives resort to get a confession from the suspect point to the inadequacy of facts alone, and also point to the larger narrative arc of Mackey's corrupt character (an issue to which I will return later). In *The Wire* individual murders are often linked together to form a larger investigation of interlocking networks of criminal activity – the facts of one case building upon the facts of another to form the links that lead the detectives up the ladder. The intimate interplay of facts on *The Wire* is taken to its absurd endpoint in the final season when Det. McNulty engages in the full-scale construction of a fictional serial killer (out of a series of unrelated deaths) in order to generate time and resources for a larger investigation into a drug ring responsible for dozens of murders in Baltimore. In one of the more humorous moments of the series, McNulty is caught in a game of Liar's Poker with a journalist who is also making up facts for his stories. In the moral world of *The Wire*, the malleability of the truth can serve either noble or disreputable purposes. Finally, on *CSI* facts are everything. As the series characters are fond of saying, "The evidence doesn't lie." But this doesn't mean that the representation of the facts is unproblematic. Instead, the facts are full of misleading details and can be interpreted in a number of ways, often depicted through competing flashbacks of the same event. Even the vaunted "*CSI*-shots" often present false details. The difference between this ambiguity on *CSI* and the way it is handled on *The Shield* and *The Wire* is that in the former series the evidence will always win out and lead to a clear solution: it just has to be *interpreted correctly*. Perhaps the key to understanding the differences among these series with regard to the status of facts and evidence is as a matter of scale and scope. While *CSI* focuses, quite literally, on the microscopic scale of the case at hand, *The Shield* builds outward from individual cases in order to see patterns of corruption within the police department. *The Wire* stretches even farther to locate patterns of corruption across all institutions and as high up in the political and organizational hierarchy as one can imagine.

A third level of "truth" in each series has to do specifically with this notion of corruption: where it exists and what to do about it. In order to get at the distinctions among the three series, I offer three brief case studies below.

The Shield: Moral Exchanges and the Return of the Repressed

In an episode from the sixth season of FX's *The Shield*, Capt. Claudette Wyms tells her former partner, Dutch: "The truth may not always lead us down the path we want, but it's the only way to fix this place." The idea of the truth as a central but not always desirable path to salvation animates the series as a whole. Like *Hill Street Blues*, *Homicide*, and *NYPD Blue*, the producers of *The Shield* created a gritty world in which crime is not so easily contained, where corruption fuels the bureaucracy of crime control, and where cops often rely on questionable tactics to make arrests and elicit confessions. But *The Shield* took this grittiness to a new level. Corruption in *The Shield* is widespread and deep. Rather than being limited to a small handful of secondary characters that pose problems for the real heroes, it is the main characters themselves who are the most corrupt. Specifically, Det. Vic Mackey and his "Strike Team," ostensibly charged with battling the gang and drug operations of Northwestern Los Angeles, are deeply invested and involved in those illegal operations, and willing to commit murder to maintain their powerful position. At the same time, the "Strike Team" consists of incredibly effective police officers who produce the kinds of arrest numbers that mayors and members of the City Council like to show to their constituents.

The series is set in the fictional "Farmington" section of Los Angeles, which stands in for the Rampart section of Los Angeles that had been shaken by corruption scandals in the 1990s involving the "CRASH" (Community Resources Against Street Hoodlums) anti-gang unit of the LAPD. Based loosely on these scandals, *The Shield* offers a serial narrative that follows the long trajectory of Vic Mackey's fall from power both within the LAPD and on the streets. Mackey's key adversaries include a series of Captains, notably David Aceveda who has designs on political office and Claudette Wyms who is dedicated to cleaning up the squad, as well as fellow detectives (especially "Dutch" Wagenbach, Wyms's partner and the cerebral counterweight to Mackey's violent masculinity), and a slew of drug dealers and gang leaders who are threatened by Mackey's criminal ambitions and are subjected to his power as a law enforcement official. At its core the series presents an elaborate and high-stakes cat-and-mouse game, with Mackey trying to stay one step ahead of both sets of adversaries: the drug lords/gang leaders whom he is trying to rule, and the police administration that is trying to control him and, in some cases, expose his corruption. At the same time, though, the police administration needs Mackey and his team for political purposes, and is mostly willing to look the other way when it is beneficial to them. Given this structure, a central thematic concern of the series is the exchange value of the truth in a mostly corrupt moral economy comprised of competing agendas.

Truth in *The Shield* acts as a kind of currency in a moral exchange about justice. The exchange value of the truth is a key ingredient in the way the series confronts the problem of corruption. Certainly, there are events and people that can be identified as having been sullied by compromise and corruption. But these compromises (professional, moral, or otherwise) animate a set of larger questions about what constitutes justice in the bigger picture. *The Shield* is populated by characters caught between their ambitions (professional, economic, and political) and their sense of what is "right." A particularly clear example of such a compromise occurs in the aforementioned pilot episode when Detectives Wagenbach and Wyms fail to get a confession from a suspected child molester. As a last resort, they involve Det. Mackey who enters the interrogation room and informs the suspect: "Good cop and bad cop just left. I'm a different kind of cop." Mackey proceeds to get the confession through violence, and while neither of the two earlier detectives are necessarily pleased with having to employ Mackey in this way, they find themselves having to compromise what they know is right in terms of procedure and the rights of the accused in order to get to the truth in the form of a confession from a guilty man.[12]

The moral compass of *The Shield* is certainly complicated, but the series is not without a certain kind of moral authority. As the passage that began this section suggests, Detective Wyms acts as the moral voice for the series, reminding other characters time and again of the importance of getting to the truth no matter how unsavory it may be. For example, in the episode "Trial By Fire" (season six), Wyms, now the precinct Captain, confronts her former partner, Dutch, who is looking for another shortcut to bring Mackey to justice. Dutch tells Wyms: "I know he's a bad guy. The evidence is the evidence." Wyms responds: "It's tempting to let sleeping dogs lie, especially when the dog's as rabid as Vic. But I've got a chance to right the ship, Dutch. To do it I've got to put an end to a pattern of silent deals and moral exchanges that have steered this place off course. That means to pursue the truth and the truth only." While this is but one example, the series is replete with conversations in which Wyms confronts fellow officers, victims, witnesses, and suspects with the need to come to the truth. What Wyms responds to consistently is the pattern of subterfuge that fuels nearly every action in the series, particularly as they concern Mackey and the Strike Team.

Subterfuge is, of course, a key ingredient of good drama, creating conflict within and between characters and enticing viewers to play along. In *The Shield* there are three specific acts that Mackey and his team try to hide, that provide the dramatic arc of the series as a whole, and produce the primary motivations for the characters. The first of these is Mackey's murder of Det. Terry Crowley in the first episode of the series. Mackey identified Crowley as working undercover to try and bring down the Strike Team and subsequently shoots him point blank in front of the other members of the team who now must remain silent or risk arrest and prison themselves. The second incident is the robbery by the Strike Team of a "money train" run by an Armenian gang. The result of this heist is a series of increasingly tenuous maneuvers to try and hide the evidence of the theft and protect the team members from revenge at the hands of the Armenian gang. They accomplish this largely through a string of lies that are designed to pit rival gangs against each other. Finally, the third incident involves a second murder within the Strike Team when Shane Vendrell kills fellow team member, Curtis Lemansky, in an effort to keep Lemansky from testifying against the team about the money-train robbery. While these incidents constitute the primary threads of the basic dramatic action across the seven seasons of the series, they are also fodder for much of the underlying emotional responses of characters to their situations: responses that reveal the various ways in which the series presents the idea of "corruption."

These incidents raise the narrative question of how the characters will respond on two levels. On one level, the response is a matter of plot: What will happen next? On another related level, the response is a matter of effect: What effect will living with these truths have on the characters themselves? Truth in *The Shield* acts something like a repressed memory. It lurks below the surfaces of nearly every move the characters make and provides the key motivation for their actions. In some cases, the truth is so painful, its subterfuge so corrupting, that it literally eats away at characters. For example Shane Vendrell, Mackey's closest ally on the Strike Team for much of the series, is so shaken by his decision to murder Lemansky and his need to hide the truth that he becomes unable to perform his role on the Strike Team effectively and eventually breaks from Mackey and the team. In a scene from the episode "Back To One" (season six), Mackey corners a gang member, Guardo, whom he suspects of Lemansky's death. Mackey and Vendrell take Guardo to an abandoned house and tie his hands to the ceiling so that Mackey can force information out of him:

MACKEY: How'd you find my guy? Was it Emelia? Cavenaugh? How!? (He hits Guardo with a chain). I'm gonna get to the truth!

GUARDO: I didn't do it! (Mackey hits Guardo with the chain again as Vendrell walks away with his hand over his mouth, horrified).

At this point, Vendrell begs Mackey not to kill Guardo. "Look, you kill him, you cut off our chance of ever finding the truth. I want every single person who had a hand in this to suffer, not just Guardo." On the one hand, Vendrell is simply acting out of self-preservation, trying to throw Mackey off the scent of his own culpability. On the other hand, his desire to save Guardo speaks to his own growing self-loathing; he cannot bear to allow Guardo to suffer alone for what he has done. Vendrell's own suffering at the hands of this hidden truth also causes a rift in his relationship with his wife, Mara, which is only cured when he confesses to her. Once Vendrell has confessed to Mara, he is willing to reveal all to his superiors. As he tells Wyms in "The Math of the Wrath" (season six): "Look, I have been lying, covering up, bleeding out of my goddamn anus, man, for three years trying to protect him. If you ever get tired of Vic walking you around on a leash and you want to hear the real and the raw, call me."

Of course, the dramatic arc of the series doesn't allow for such an easy path to Mackey's demise. In the final season, Mackey is finally cornered but uses the vast information he has to get immunity for himself. Vendrell ends up killing himself in light of the recognition that Mackey has manipulated the system and is able to confess all his misdeeds, but receive full immunity and no jail time and, seemingly, feel no remorse for what he has done. While the moral authority that Wyms represents – with the truth as a path to salvation – guides much of the series, Mackey's ultimate victory (avoiding jail time) suggests that truth is also a commodity of corruption that can be used strategically to buy and maintain one's position. The end of Mackey's reign at Farmington is a double-edged sword for the audience. The truth has been revealed and the corruption has been contained for now; but we've been reveling in that corruption through the charismatic character of

Mackey for seven seasons. His demise is both narratively satisfying and emotionally disheartening. On the other hand, Mackey has managed to use the truth to his own advantage to secure his freedom and not pay fully for his corruption; but at the same time, his freedom is a monotonous sentence of its own kind, highlighted in his last scene by the steady buzzing of the overhead fluorescent lights that illuminate his beige cubicle. The truth will set you free. Sort of.

The Wire: The Sociological Imagination of Crime

If *The Shield* represents the truth as a possible path to salvation, a way out of corruption, HBO's *The Wire* illuminates the underlying truth *of* corruption: that it is widespread and tied indelibly to the structure of our social, economic, and political institutions. *The Wire* takes Lord Acton's observation that "absolute power corrupts absolutely" as its guiding vision to demonstrate the ways in which every aspect of our social lives is defined by unequal power relationships – whether that be on the streets or in the highest echelons of political office. The series ties these notions of power and corruption to a string of intersecting narratives that attempt to illuminate certain truths about life in urban America in the 21st century. The stories that *The Wire* tells about truth, then, are not necessarily about the facts of a particular case – though that element of police work is vital to the dramatic work of the series. Instead, the series engages with what we might call the "sociological imagination" of crime and corruption in urban America.[13] That is, the series works to construct a more interconnected sense of how the specific stories of crime can point to larger social and political issues.

The "sociological imagination" is a concept developed most forcefully by sociologist C. Wright Mills in 1959 in an effort to push the social sciences toward a larger view of the intersections between individual experience and social life. For Mills, the "sociological imagination" was comprised in the ability to "range from the most impersonal and remote transformations to the most intimate features of the human self – and to see the relations between the two" (Mills 1959: 7). This is precisely the work of *The Wire*: to engage with the immediate stories of drug dealers, homicide detectives, inner city educators, dock workers, politicians, journalists, and everyone else in between in order to address the ways in which we are all caught up in the institutional power struggles that guide the minutiae of our daily lives.

Tellingly, *The Wire* begins with two scenes that illustrate this interplay nicely. In the opening scene of the series, Det. McNulty is trying to get information about a dead body lying in the street. He is told that the deceased is a boy nicknamed "Snotboogie" who had a habit of joining dice games in the alleyways and, at some point, stealing the money and running away. According to McNulty's witness "Snot" would typically get beaten up for this indiscretion. This time, he got shot and killed. McNulty seems confused. "I gotta ask you. If every time Snotboogie would grab the money and run away, why'd you even let him in the game?" McNulty's witness responds simply: "Got to. It's America, man." This basic scene illustrates one of the key strategies of the series: connecting the individual to the social. As Rafael Alvarez puts it: "The Snotboogies of the world may be comical and doomed and they are often dangerous. But as citizens of a parallel world under the same Stars and Stripes as the legitimate America, they must be allowed to play" (2009: 64). The story of Snotboogie is not only the story of a single "corner boy" who winds up dead; it's the story of two Americas with parallel structures and economies, in which everyone gets to play regardless of how thoroughly the game is fixed.

The next sequence depicts a courtroom trial in which a witness suddenly reverses her testimony under threat from the Barksdale crime family. McNulty notices one of the members of the family, "Stringer" Bell, in the courtroom and realizes that intimidation is at the heart of this sudden reversal. McNulty is called into chambers by the presiding judge and reveals that the Barksdale family is intimidating witnesses at every turn and beating an otherwise airtight prosecution. Disgusted, the judge asks McNulty who in the department is working on the Barksdale family. McNulty tells him that nobody is. At this point, the judge paves the way for the establishment of a special team to work on the Barksdale case. I've oversimplified the dynamics of this scene and the scenario it sets up, but the larger point is that the judge (and the series creators) establish early on a relatively simple solution to a social problem: reallocation of resources within the police department to address a real threat to the community. But simple solutions get tangled in webs of politics and corruption. The team given to the judge's project by the Baltimore Police Department is comprised mostly of officers who are no longer useful to their divisions: a team designed to fail. While the judge would like to damage the higher levels of the Barksdale operation, the police department, under pressure from its higher-level administration and the mayor's office, is more concerned with numbers gleaned from smaller street-level arrests of small-time players. These are the politics of police work that rely more on tangible numbers and "juking the stats" to make the department look successful despite the lack of any real progress in stemming the flow of drugs and violence throughout the city. In spite of these constraints, however, the team begins to gather information (via wiretaps that give the series its title) and begin to string important pieces of information together that might help them do some damage to the Barskdale operation, but they are thwarted at every turn by an opportunistic and sometimes corrupt power structure that needs "all the pieces" to stay in place in order to maintain a particular hierarchy of power.

A similar situation arises in the third season of the series when Major Howard "Bunny" Colvin develops a radical plan to make inroads into the spread of drugs across his precinct. Rather than try to fight a diffuse and ultimately ineffectual war on the corners, Colvin offers the players in the drug trade basic immunity in return for their relocation to a nearly abandoned part of the city that becomes known as "Hamsterdam." The idea from Colvin's perspective is to contain the drug trade within a clearly defined area that can be more easily monitored and controlled. The series establishes Colvin's ostensible decriminalization of drugs as a potentially successful, if ultimately doomed, experiment in restoring the life and vitality to the inner city environment that too many politicians have left for dead. As Peter Clandfield points out: "The environmental implications of Colvin's scheme are highlighted by the way the camera repeatedly frames him against backdrops of green vegetation" (2009: 42) when he is within the now drug-free streets of his precinct. The evidence on screen suggests that Colvin's plan is making a difference. But his tour of the area (both the rejuvenated neighborhoods and "Hamsterdam") with mayoral candidate Tommie Carcetti presages the general reaction of Colvin's superiors and the media who leap upon the political and economically expedient opportunities of denouncing the "legalization of drugs" in any way, shape, or form, regardless of the progress (seen in what the camera shows us and what the crime statistics that Colvin reports tell us). Once wind of the project reaches the media and an outraged public, Colvin is forced to shut down the operation and is, in fact, eventually forced to resign from the department. *The Wire*, then, is working on two levels simultaneously. On the one hand, the series creates a recognizable crime story through the very particular cases involving the Barksdale family specifically, and the war on drugs more generally. At the same time, though, the series allows viewers to share the frustration of the police detectives as they begin to make larger connections that are mostly ignored – or even punished – by their superiors.

In fact, the audience is encouraged to make even broader connections between the crime story at hand and the socio-economic and political contexts within which this criminal activity takes place. One particular trope of the series is to craft conversations between characters that act somewhat like a "Greek Chorus" for the series. Consider the following example from the episode "The Detail" (season one) in which three young dealers in the housing projects discuss Chicken McNuggets:

WALLACE: Man, whoever invented these, he off the hook!

POOT: What?

WALLACE: Motherfucker got the bone out all the damn chicken. Till he came along niggers been chewing on drumsticks and shit, getting their fingers all greasy. He said, "Leave the bone. Snug that meat up and get some real money."

POOT: You think the man got paid?

WALLACE: Shit, he richer than a motherfucker.

D'ANGELO: Why? You think he get a percentage?

WALLACE: Why not?

D'ANGELO: Nigger, please. The man who invented them things is just some sad ass down at the basement at McDonald's, thinking up some shit to make some money for the real players.

POOT: No, man, that ain't right.

D'ANGELO: Fuck right. It ain't about right, it's about money. Now you think Ronald McDonald gonna go down that basement and say, "Hey, Mr. Nugget. You the bomb. We selling chicken faster than you can tear the bone out. So I'm a write my clowny-ass name on this fat-ass check for you." Man, the nigger who invented them things – still working in the basement for regular wage thinking of some shit to make the fries taste better.

WALLACE: Still had the idea, though.

While the conversation works on one level to establish a realistic tone and mood for the series characters, establish their relationship with one another, and providing some comic relief, it also works to emphasize a socio-economic truth about the exploitative labor relations that have defined American capitalism – relations that define the daily lives of these small players in the drug trade and that provide the real impetus for the series. In the words of C. Wright Mills: "Perhaps the most fruitful distinction with which the sociological imagination works is between 'the personal troubles of the milieu' and 'the public issues of social structure.'" The "troubles" of which Mills writes are private matters that lie within the individual and that "have to do with his self and with those limited areas of social life of which he is directly and personally aware." Meanwhile, "issues" are more public, social matters: "They have to do with the organization of many such milieux into the institutions of an historical society as a whole, with the ways in which various milieux overlap and interpenetrate to form the larger structure of social and historical life" (1959: 8). What is at stake in this seemingly innocuous conversation about Chicken McNuggets is the complex interplay between the situational "troubles" of the three characters and the social "issues" that circumscribe their existence.

Similarly, in a scene from the next episode, "The Buys" (season one), D'Angelo explains the game of chess to two of his drug runners: Wallace and Brodie. The explanation quickly turns into a clear discourse on the structure of the drug trade, with Wallace and Brodie commenting overtly on the similarities between the pieces and their drug-trade counterparts. Brodie says the Queen "Reminds me of Stringer" when D'Angelo describes the piece as the "get shit done" player. The scene also stands as a description of the larger social structure outside the drug trade as well. The fact that D'Angelo is describing a game resonates with an oft repeated phrase in the series: "All in the game," meaning both that the rules are what they are, and on another level, that there are few distinctions between "legitimate" and "illegal" institutional practices. In the episode "One Arrest" (season one), Omar Little, a violent and charismatic independent player in the drug trade, who makes his living robbing drug dealers, finds himself testifying in court against one of the Barksdale crew members for killing a witness. The defense lawyer, Maurice Levy, tries to discredit Omar as just the kind of killer he is claiming his client to be:

LEVY: You are amoral are you not? You are feeding off the violence and despair of the drug trade. You're stealing from those who themselves are stealing the lifeblood from our city. You are a parasite who leeches off…

OMAR: Just like you.
LEVY: ...the culture of drugs...excuse me? What?
OMAR: I got the shotgun. You got the briefcase. It's all in the game, though. Right?

The point is clear: while on the surface Omar and Levy couldn't be more different, they are both players in a larger game. Importantly, it's not just that they are both involved (though at different levels) in the drug trade (Levy defends drug dealers in court), but that they are representatives of the same impulse toward self-preservation in a world that, according to the creators of *The Wire*, doesn't reward honesty and individual sacrifice. For Mills, the key for both Omar and Levy to make sense of their "troubles" would be to take a broader view of the "issues" – to understand the ways in which they are interconnected into a larger social structure. It is precisely this work that *The Wire* undertakes with regard to illuminating the truth of corruption in 21st century urban America.

I have focused mostly on the first season of *The Wire* because it is within that season that the major themes of the series – regarding institutional corruption at all levels – are set in motion. The remaining seasons deal with corruption within a range of different institutions: the ports of entry (season two), politics (season three), education (season four), and the media (season five). While these later seasons continue to build on the criminal investigation of the drug trade that comprises the long arc of the series, each season delves into its particular institutional setting with the same eye toward crafting connections between individuals and the institutions that guide their daily lives. In the end, the truths of *The Wire* transcend its specific location in Baltimore to offer a story about corruption and the near impossibility of true reform writ large.

CSI: Nothing But the Truth?

While *The Wire* establishes a much larger political and thematic canvas than more traditional police procedurals, a good portion of the series revels in the intricacies of everyday police work, especially the more clandestine work of stakeouts and wiretapping. The series basks in the meaningful details that hardworking detectives are able to glean from tiny strands of evidence, such as a turn of phrase, or a pattern of numbers. In this way, the series connects to the traditional procedural as represented by *CSI*. As discussed earlier, *CSI* is a series built on the fetishization of visual evidence as presented to the viewer through flashbacks and, most memorably, through CGI presentations of the human body in various states of distress. As opposed to the notion of truth as personal salvation (*The Shield*) or truth as a product of the "sociological imagination" (*The Wire*), however, *CSI* is concerned primarily with the status of evidence as the locus of truth. As Gil Grissom, the first shift supervisor on the series (played by William Petersen), was fond of reminding his underlings: "People lie, but evidence never does."

In fact, the narrative arc of each episode typically involves sorting through the chaos of the initial mess of evidence and human interpretation in order to let the evidence speak. This chaos can take several forms: from improperly stored evidence, to personal feelings, to unreliable witnesses. As Grissom tells CSI Warrick Brown in "Friends and Lovers" (season one) when Brown asks Grissom how he feels about the crime they are investigating: "It doesn't matter how I feel. Evidence only knows one thing: the truth. It is what it is." Personal interpretation is not only beside the point; it is potentially disruptive to reaching the actual truth. Similarly, when CSI Sarah Sidle allows herself to get caught up in finding out about the life of a still living but comatose victim ("Too Tough To Die," season one), Grissom advises her to get a hobby outside of the job – the idea being that allowing personal feelings into the job will result in either burnout, or misreading evidence, or both. A middle ground exists, however, in the character of CSI Catherine Willow who counterbalances Grissom's reluctance to engage in the human elements of the crime. In "Friends and Lovers" (season one), Willow distances herself from Grissom: "This is where Grissom and I differ. Forensics is about more than science. Human behavior, the inconsistencies of human behavior. His 'how' is crucially important. But so is 'why'." What emerges from these examples, then, is a range (though an admittedly narrow range) of discourses about how to arrive at the truth. The series certainly privileges scientific discourses about observable evidence, played out through its visual style, but also opens up a space for the more humanistic discourses about motives and "the inconsistencies of human behavior."

The issue of "human behavior," in fact, is often at play across the series in the way characters respond to the murders and evidence – ways that are not so easily trumped by science. For example, in "Friends and Lovers" (season one), the question of *why* the murder occurred competes with the typical and central question of *how*. As with most *CSI* episodes, "Friends and Lovers" is built around two primary narrative threads. The first involves the dean of a private school who has been killed by a pair of female teachers who are in a romantic relationship and being blackmailed by the dean. The second involves a young man found dead and naked in the desert. It turns out that he was killed by a friend who was having auditory hallucinations after drinking jimson tea and was trying to keep his friend (who was also hallucinating and frightened) quiet. In keeping with the basic aesthetic and narrative structure of the series, the *how* of the crime takes center stage in terms of the presentation of visual evidence. But in both of these cases the motivations behind the murders (the politics of sexuality in one case, and a simply unfortunate, unforeseeable, and utterly irrational circumstance in the other) are allowed additional narrative space and threaten to undermine the dominant scientific/forensic discourse of the series. In the case of the first murder, upon concluding their investigation (locating the *how* and *who*), CSIs Nick Stokes and Catherine Willow ponder the results:

NICK: You got your why. A crime of passion.
WILLOW: The bigger why: why did it have to come to this?
NICK: I guess they didn't feel like they had a choice.
WILLOW: Maybe they didn't. A lose, lose situation.

Similarly, when confronted with the evidence of his role in his friend's death, but also of his relative innocence (he was quite literally not in his right mind), the young man in Grissom's case is wracked by guilt and cannot forgive himself. Grissom shares his disappointment with the disparity between the legal truth and the "justice" of arresting this young man. In both cases, the *how* and the *why* point in opposite directions in terms of guilt and/or culpability. While *CSI* often constructs narratives that neatly overlay the "guilty act" and the "guilty mind" – as the evidence of one set of facts unproblematically asserts the other – these two examples suggest that the two cannot always be so easily conflated: that human motivation is a more complex set of factors than any evidence can suggest.

Mostly, however, the complexity of human behavior is seen as a problem to be overcome, particularly as it involves witnesses who are seen as largely unreliable in comparison to the processes of scientific discovery. In "Unfriendly Skies" (season one) the mysterious death of a passenger on a commercial flight yields a range of competing witness accounts, each of which clouds the investigation in one way or another. This narrative trope echoes the structure of the classical detective story in which a parade of suspects and witnesses present equally compelling (and equally suspicious) accounts of what may or may not have happened until the detective finally arrives at a solution that is different from all previous stories.[14] In this case, though, it's not the superior intellect of the detective, but the superior (i.e. disinterested) perspective of scientific method that allows the final solution to emerge. In the end, several of the passengers grouped together to kill the victim believing his erratic behavior (due to encephalitis) to be a threat to their own safety. Importantly, Grissom acknowledges at the end of the episode that had the other passengers simply asked the victim what was wrong, instead of making assumptions about what his behavior meant, his death might have been avoided. In this case, Grissom recognizes the human element as both problematic in terms of the investigation (due to inherent speculation and erroneous interpretation) and also the key to avoiding the victim's demise in the first place (through acts of simple interpersonal kindness).

Witnesses are not always deceptive as much as simply wrong. In "Too Tough To Die" (season one), Willow and Brown reopen an older investigation and find that the key witness in the case (the wife of the victim) could not possibly have seen what she claims to have seen. Instead, she has simply filled in the gaps of what she imagines must have happened before she arrived. Similarly, in "Who Are You" (season one), the investigation of a shooting involving a police officer yields a witness who claims to have witnessed the incident first hand. He tells the CSIs that the officer in question shot the victim without provocation, contradicting the officer's testimony that the victim shot himself. Once again, a careful investigation of the evidence proves the officer to be telling the truth; the witness simply extrapolated from incomplete evidence what he *thought* he saw. While the act of extrapolating from incomplete information and filling in the gaps is a crucial and unavoidable activity of readers and viewers of narratives, in the case of *CSI*'s narrative world, it is an activity that needs to be tempered by science.

The narrative and aesthetic treatment of misleading testimony in *CSI* suggests on the one hand that "truth" is a matter of perspective: the witnesses are convinced of the truth of their accounts. But, like most narratives, the series works hard to establish a hierarchy of knowledge that privileges one set of perspectives over another. In the case of *CSI*, as discussed earlier, the visual retains the highest level of authority – a fact borne out by the visual style of the show which fetishizes intimate details via the "*CSI*-shot" – in order to underscore the veracity of the evidence at hand.[15] In particular, we might refer to the privileged position of ocular perspective as constructed through both the CGI shots of the human body (what the eye cannot truly see) as well as the frequent macro zooms to evidence at the crime scenes, which are designed to emulate the visual perspective or literal viewpoint of a particular character (always one of the investigators). Though this simulated visual perspective is impossible for the human eye in terms of microscopic detail, it crucially connects the human investigator with the forensic equipment he/she uses in the lab. Thus, the enhanced visual abilities of scientific instrumentation are displaced onto the human figure (particularly the human eye), which is then granted the authority of science itself.

Importantly, though, the characters are not simply reduced to an equation with scientific technologies. Instead, a second level of knowledge is also offered up as an important tool in the investigators' kit. This second level is that of human experience itself. For example, in the episode "Too Tough To Die" (season one), the evidence in the shooting that Brown and Willow are investigating continues to baffle them as they refer to the coroner's charts for entry and exit wounds from a gunshot. The figures in the charts are constructed in traditional fashion to depict the human body (the body of the victim) standing erect with arms at the side, slightly elevated from the body. But Brown keeps returning to the fact that the victim and the suspect were fighting with each other at the time of the shooting. "Guys don't fight like gingerbread men," says Brown. He then simulates what a real fight between two men would look like: lots of grappling and body contortions. Only by removing themselves from the clinical/scientific diagram of the victim's wounds, and moving into the real, bodily world of human interaction can Brown and Willow arrive at the right answer. The narrative structure and visual style of the series allow competing perspectives to play out, but at the same time the series establishes a clear hierarchy of knowledge that privileges scientific discourses and knowledge based on direct experience to trump the emotional testimony of witnesses and suspects.

What the three case studies presented here illustrate is the degree to which each of these police dramas trades in representations of corruption (or the total lack of it) that accrue around a larger question of how one knows and uses "truth." In *The Shield*, the truth is both a currency to be used in the service of personal and professional ambitions as well as a path to salvation. In *The Wire*, both the concept of truth and the problem of corruption are understood broadly – as matters of interlocking cultural, social, political, and economic forces that cannot be easily contained by the work of the police. In the world of *CSI*, the only corruption of any consequence is the corruption of evidence that can lead away from the truth. This corruption can take any number of forms, from misleading testimony, to mishandling of materials, to allowing one's self to become tainted by personal responses to the crimes. As Gil Grissom tells Brown in "Too Tough To Die" (season one): "We don't impose our hopes on the evidence."

Conclusion

The police dramas of the 21st century constitute a development in the cultural forum on crime, community, and citizenship. On the one hand, they continue to elaborate upon these discourses through the stories they tell. As the case studies presented here suggest, issues of truth and corruption correspond to discourses about crime, community, and citizenship in several ways: how we know and identify crime and justice in the first place through evidence and testimony; the effects of corruption on the economic, political, and social functioning of the communities within which we all live; and the responsibilities and obligations we have as citizens to truth in the pursuit of justice. At the same time, these series present very different approaches to these questions, largely due to their respective positions within a changing media environment that encourages stories designed for increasingly differentiated and narrow slices of the viewing audience. The "forum" these series participate in, then, is more diffuse than ever in terms of both form and ideology.

These series operate on a number of levels at once. First and foremost, they are products of an industry in rapid transition, in which the very concept of television itself is in play, and in which the value of the "brand" is paramount to the logic of the industry. Understanding the series in question as "branded" in specific ways allows us to more specifically locate the enabling conditions of their production and perhaps more accurately analyze their different sensibilities. These sensibilities emerge from a range of contexts: institutional locations, regulatory frameworks, suppositions about audience demographics, and expectations of quality. In other words, understanding how these series are positioned in relation to specific brands (networks, production companies, etc.) goes a long way toward allowing us to explain how they function as both quite distinct, but also generically recognizable, texts.

What a focus on branding may in fact encourage is a critical reconsideration of genre identity. As Jason Mittell has argued, genre identity is less a matter of particular textual features that may or may not "evolve" over time, and more a matter of the "cultural life" of a category, which privileges some characteristics over others at different times and under different institutional, popular, and critical conditions.[16] What the contemporary moment may suggest is that the "police" genre may no longer matter as much as the specific brand identity to which a particular text is attached. For instance, perhaps it is more fruitful to think of FX programs as a more compelling genre than "police dramas." Series like *The Shield* may have more in common with other FX series, such as *Justified* or *Sons of Anarchy* than with *CSI* or even *The Wire.* Additionally, as series move across networks in syndication, and connect to other series procured for a particular network brand, their identity may shift again.

In the end however, athough *CSI*, *The Shield*, and *The Wire* represent very different sensibilities with regard to how they tell stories about crime, and were produced for different programming platforms, there is no necessary correlation between the programming outlet and the ideological position of the series. In other words, there is nothing about the relative freedom with regard to content that HBO enjoys that guarantees a more "progressive" politics. Similarly, nothing about the formulaic structure of *CSI*, which has served CBS so well, guarantees a more "conservative" sensibility. In fact, is has been my argument all along that police dramas as a whole cover a range of ideological positions both within and across series. What this means, quite simply, is that the genre is capable of very different sensibilities at any given time, under any given economic, social, or political circumstances.

8

CONCLUSION

The police genre has experienced a rebirth of sorts during the last three decades within a constellation of institutional practices, technological developments, stylistic experimentation, and social discourses in transition. This period has seen the fully-fledged emergence of cable and satellite distribution as popular alternatives to network broadcasting, and the emergence of web-based distribution as a key platform in the further transformation of the media industries writ large. These new technologies have significantly expanded the range of programming available to viewers. Increased competition for audiences, spurred by these technologies, has encouraged the networks to seek programming that can compete for portions of the audience that have been dropping away in favor of ad-supported cable networks like FX, Spike, TNT, and USA, as well as subscription services like HBO: especially young men and higher-income professionals. New programming strategies are also being encouraged by advertisers seeking to maximize their investments by more efficiently targeting narrower segments of the potential viewing audience. Television programmers have responded by offering more programs designed to attract narrower segments of the "quality" audience (those with higher-salaries and more expendable income). Targeting these audiences requires programs that can stand out amidst the clutter. In order to achieve this differentiation, producers pull from a wide range of cultural sources. Documentary and feature films, pop art and advertising, reality television, and soap operas, for example, provide an array of stylistic and narrative possibilities from which to choose.

This period is also marked by social, political, and economic developments that have had an impact on the police genre. The decline and then gentrification of inner cities across the country has fueled an uneven crackdown on street crime, especially drugs, while largely ignoring the larger social and economic contexts in which these changes have taken place. Tighter restrictions on the police in terms of their ability to arrest and interrogate suspects, as well as their ability to gather and introduce evidence, has fostered a reactionary victims' rights movement that changed from being concerned with assuring the fair treatment of minority and female victims to a focus on retribution against perpetrators. Increased complaints about police corruption and systemic racism within the police continues to drive a wedge between them and many minority groups. Finally, the rise of community policing initiatives continues to open up questions about the limits of citizens action as well as concerns about creating harmful boundaries between communities, especially based on race and class.

Given this array of issues, I have argued that the cycle of police dramas that emerged since 1981 has constituted a rich cultural forum around the issues of crime, community, and citizenship. These themes have always been at the center of police dramas, both fictional and non-fictional, from *Dragnet* to *COPS*, in television, film, and literature. But they are not static concepts. Rather, they are *social* and *discursive* concepts; they are only meaningful within the context of fluid social relationships and interactions. They are also *political* and *ideological* ideas, subject to struggle over their definitions. This study has been an attempt to isolate these ideas within a particular historical moment, and within a specific generic and institutional context, in order to examine them in action and to offer some insights into how cultural institutions such as commercial television interpret social ideas and circulate notions about them.

To speak of a cycle of television dramas as a cultural forum highlights the ways that television narratives open up spaces for reflection about social and cultural concerns. As Newcomb and Hirsch argue, "television does not present firm ideological conclusions – despite its *formal* conclusions – so much as it *comments on* ideological problems" (1983: 506). This idea runs counter to much of the criticism that has been made of the police genre. In particular, the police genre has been seen as a hopelessly conservative form that offers overly simplistic solutions to the complex problem of justice. But these criticisms are focused on the formal conclusions of the narratives. Newcomb and Hirsch's model, on the other hand, asks us to look more closely at the range of issues that are put into play within specific narratives, across series, genres, and the schedule as a whole.

The cultural forum model itself must be understood in context. Newcomb and Hirsch were writing at the tail end of the network era (1983), just before cable and satellite technologies significantly expanded the television landscape. At this time, the big three networks still dominated the market. As Amanda Lotz has demonstrated, an important foundation for the cultural forum model was the issue of television's "scope": its massive reach during a period of relatively limited choice created something like a shared experience (2004: 25). The continued rapid growth of the television industry in the period since Newcomb and Hirsch's article has made this issue of scope less applicable than it was over twenty years ago. Viewers are more dispersed than ever before, and are likely to simply ignore programming that they find troubling or offensive. Additionally, producers and programmers are no longer invested in always attracting a truly mass audience. Niche markets built around specific brands have proven to be extraordinarily profitable. Given this institutional reorganization, we need to rethink the cultural forum model.

Newcomb and Hirsch argued that critical analyses based on the cultural forum model should focus on "strips" of programming in order to approximate the actual viewing experience: the flow of programming. These strips were likely to feature very different types of programming flowing into one another: sitcoms into dramas into news for example. The range of program types in this flow would necessarily contain a wide range of "assumptions about who and what we are" (Newcomb and Hirsch 1983: 508), and these competing assumptions would come into contact with one another at the level of the viewer. While I believe that this is still a vitally important method to follow, the number of potential viewing strips has expanded so greatly in the past two decades that it is now possible to imagine entire strips comprised of similar types of programs. FX, for instance, is devoted to programming designed primarily for a young male demographic, and has developed a range of original dramas that bear distinct similarities to one another in terms of the attitude they project. Thus, the viewing strip on FX is far less differentiated than a strip on ABC during the same timeframe. For this reason, I have argued that the cultural forum model can be narrowed and reconstituted as a quality of genres.

Within the police genre, I have focused on a number of series. Some of these series are considered more traditional and thus conservative (such as *T.J. Hooker*, *Hunter*, and *CSI*). Others are considered to be innovative in terms of their style and their narrative complexity, as well as being more progressive politically (*Hill Street Blues*, *Homicide*, *NYPD Blue*, *Law & Order*, *The Wire*, and *The Shield*). What I hope to have demonstrated is that the discourses of crime, community, and citizenship are central to all of these series, despite their perceived quality. I have not been interested in determining which series are "better" or more politically savory than others. Rather, my point has been to demonstrate how these discursive issues get played out across and even *within* particular series. I have tried to show how these series exist in a dialogic relationship to one another. Rather than focusing on a strict ideological unity, either within the genre as a whole, or within any given series, I am interested in the "struggle for meaning" that is at the heart of our social relationships and which gets played out nightly on our television screens.

My analysis has relied in part on case studies that focus on individual discourses within discrete series. This isolation of single series and particular thematic concerns is in no way intended to suggest that mine is the only way to read the series in question, or that the series can be read only through the lens of that particular discourse. In fact, all the series examined here are invested in each of these discourses, and more. Future analysis would do well to open up different configurations and discursive struggles.

This cycle of the police genre encompasses the events of September 11, 2001, a point I have left largely unexplored. While my initial impulse to bracket off 9/11 in my analyses had much to do with wanting to explore changes in the industry, especially the importance of franchising series and branding, as well as a concern that such a focus would necessarily override my more immediate critical concerns about the enabling conditions of the television industry in the U.S. But bracketing 9/11 doesn't foreclose the kinds of questions that future research might ask. In particular, how have our notions of crime, community, and citizenship changed since 9/11? How have social debates about terrorism, the war on drugs, civil rights, and victims' rights shifted in this new period? How has the symbolic relationship of the police officer to heroic action changed in light of 9/11? How have attitudes about authority changed in police dramas? What kinds of plots seem off limits in the current political climate?

As social discourse changes, and the thematic stakes of the police drama change with them, we should continue to ask how these series continue to constitute a cultural forum, and how television, more generally, continues to help shape our understanding of the world around us. Television, even from its earliest days, has always presented at least two very different faces: friend and foe. As our culture's central storyteller, it is at once Newton Minow's "vast wasteland" and the glowing hearth around which we all gather: benevolent teacher, bad influence, annoying guest, trusted friend. And as a cultural force television has always moved in two opposing directions: outward and inward. The centrifugal push of television as an increasingly sophisticated web of technologies opens a larger and larger "window on the world," allowing us to travel (with increasing immediacy) to distant and foreign places – to see "others" both up close and at a distance. The dominance that was enjoyed by the three networks for almost three decades came as close as anything to forging what one critic has labeled "one nation under television." But that relative coherence has been unraveled by an ever-expanding universe of specialized cable and satellite channels, internet distribution, not to mention DVR technology that allows viewers to create their own viewing schedules, and mobile technologies that allow us to access those schedules and viewing options from a distance. Meanwhile, television still pulls toward the center; the centripetal force of this commercial and mostly domestic medium centralizes information, filtering the scary world "out there" for safe consumption.

Stories about policing have always lived at the center of this tension, showing us a social system that is continuously threatened by crime, and making the successful containment of that crime a central narrative concern in the maintenance of social order. As such, stories about crime and punishment are stories about social practice and social power. If the cycle of police dramas that has emerged over the past three decades can be seen as offering something new to our understanding of the genre, our critical approaches need to move beyond strictly aesthetic concerns and, more importantly, beyond ideological axe grinding. As this study has tried to show, the police genre has never been just one thing.

Like the television schedule itself, the police drama can be seen as a forum or arena in which a range of ideas circulate and vie for our attention. Indeed, the commercial demands of network television circumscribe the range of "acceptable" ideas that are allowed to circulate. But it is important to recognize that the production and consumption of television involves a complex set of interactions involving media institutions, social institutions, and political agendas which are manifested in a variety of ways across a range of programs and finally taken up on our screens (whatever those may be) in quite personal ways.

I am reminded here of the argument put forth by David Morley and Kevin Robins about the idea of the screen as not just a place for projection, but for filtering as well:

> If in one sense, screening means that "they" are made present to "us" in representation, it is also the case that the image of "them" is screened in the different sense of being filtered, with only certain selected images getting through. At the same time, in a psychic sense, the screen is not only the medium through which images are projected for us, but also the screen onto which we project our own fears, fantasies and desires concerning the Others against whom our identities are defined and constructed.
>
> *(1995: 134)*

The police genre, of which each of the shows analyzed here is a part, becomes a vital site for investigating these projections. Concerned as it is with the dichotomies of good and evil, us and them, the police show has always worked at the borders of identity to offer certainty and protection in a chaotic and threatening world. The opening of these borders following the demise of Cold War dichotomies, combined with the increasing transnational flows of people and images, allows us to screen more of the world than ever before. The police genre is one of the sites where we make sense of that world.

NOTES

1 Introduction

1 In 2008, Court TV changed its name to TruTV and, while it has retained its primary focus on legal issues, the network has branched out into a broader array of reality programming as well.
2 See Steve Jenkins (1984), "Hill Street Blues" in *MTM: Quality Television*; Brooks Robards, (1985), "The Police Show" in *TV Genres: A Handbook and Reference Guide*; John Sumser (1996), *Morality and Social Order in the Television Crime Drama*; David Marc (1996), *Demographic Vistas*; John Fiske and John Hartley (1978), *Reading Television*; John Fiske (1987), *Television Culture*; David Buxton (1990), *From The Avengers to Miami Vice*; and Todd Gitlin (2000), *Inside Prime Time*.
3 For an excellent discussion of how a range of personal and institutional and political factors influence the development of a series, see Richard Lindheim and Richard Blum (1991), *Inside Television Producing*. Lindheim and Blum provide a detailed account of the development of *Law & Order*, from conception through the completion of the pilot.
4 In his excellent study of the development of *Hill Street Blues*, Todd Gitlin locates the success of the series at the end of a long series of flukes:

> "In network television, even the exceptions reveal the rules. Everything emerges at the end of a chain of *ifs*. *If* a producer gets on the inside track; *if* he or she has strong ideas and fights for them intelligently; *if* they appear at least somewhat compatible with the network's conventional wisdom about what a show ought to be at a particular moment; *if* the producer is willing to give ground here and there; *if* he or she is protected by a powerhouse production company that the network is loath to kick around; *if* the network has the right niche for the show; *if* the project catches the eye of the right executive at the right time, and doesn't get lost in the shuffle when the guardian executive changes jobs...then the system that cranks out mind candy occasionally proves hospitable to something else, while at the same time betraying its limits" (2000: 273).

While I do not disagree with Gitlin's colorful (and accurate) description of the vagaries of network television decision-making, his central point is that, above all, television is an abject cultural form that can only occasionally stumble into something better than itself. Gitlin's is an admirably political, but ultimately totalizing vision of how television functions as a cultural phenomenon.

5 A survey on the fear of crime, conducted between 1980 and 1982, reported that "40 percent of all Americans are highly fearful of violent crime." See Research and Forecast, Inc., and Ardy Friedberg (1983), *America Afraid*, p. 43.

6 In fact, it could be argued that, given the general lack of experience that most of us have with violent crime, the police (and the crimes they investigate) are greatly over-represented on television. By invoking this over-representation, however, I do not wish to engage in ongoing debates about violence on television or the "cultivation" of fear and anxiety about crime – a subject on which George Gerbner and his colleagues have written widely. My central concern with their possible over-representation has more to do with the simple fascination we have with images of crime and punishment, even as they outrun our own experiences.

7 Of course, the police genre was not alone in this movement toward serialization. Medical dramas such as *St. Elsewhere*, *Chicago Hope*, and *E.R.*, as well as legal dramas such as *L.A. Law* and *The Practice* all embraced the serial format. In addition, a range of other series in the 1980s and 1990s were either primarily serials (*Twin Peaks*, *Thirtysomething*, *China Beach*, *Picket Fences*, *The X-Files*, *Party of Five*, *Northern Exposure*, *Moonlighting*, *I'll Fly Away*, to name only a few), or contained at least some serial elements within a mostly episodic format (*Magnum, P.I.*, *Northern Exposure*, *Profiler*). Even situation comedies incorporated serial elements into their largely episodic narratives: *Cheers*, *Murphy Brown*, and *Roseanne* come to mind.

8 See D'Acci (1994), *Defining Women*; Butler (1985), "Miami Vice: The Legacy of Film Noir"; and Jenkins (1984), "Hill Street Blues."

9 See Joseph Turow (1997), *Breaking Up America*. Turow illustrates how the American media industries increasingly rely on lifestyle divisions in order to target specific communities with specifically designed marketing campaigns. As Turow states early on, "in the highly competitive media environment of the 1980s and 1990s, cable companies aiming to lure desirable types to specialized formats have felt the need to create 'signature' materials that both drew the 'right' people and signaled the 'wrong' people that they ought to go away" (1997: 5). These groups of "right" and "wrong" audiences form what Turow calls "primary media communities" (1997: 7). At the far end of this development, Turow imagines a world of individual consumers increasingly isolated from the bonds of community life: "market segmentation and targeting may accelerate an erosion of the tolerance and mutual dependence between diverse groups that enables society to work" (1997: 7). It is this tension between the individual and the community that I argue the police series negotiates in a range of often conflicting ways.

10 See David Marc (1996), *Demographic Vistas*, p. xxvii; see also Michael Curtin (1996), "On Edge: Culture Industries in the Neo-Network Era."

11 According to Curtin, the neo-network era is characterized by "multiple circuits of information and expression" as well as "flexible corporate frameworks" for exploiting "diverse forms of creativity toward profitable ends" (1996: 197). See also Lotz (2007b), *The Television Will Be Revolutionized*.

12 The Discovery Channel, The Learning Channel, and Animal Planet are all owned by Discovery Communications, Inc. See Cynthia Chris (2002), "All Documentary, All the Time?"

13 See John Horton (1981), "The Rise of the Right: A Global View." See also Tony Platt and Paul Takagi (1981), "Law and Order in the 1980's." Platt and Takagi state:

> "The current 'law and order' campaign, orchestrated at the highest levels of federal, state, and local governments, is well under way to eliminate the minimal reforms in criminal justice and corrections that were won in the 1970s. The justification for this shift to tougher punishments – deterrence, incapacitation, mandatory sentences, restitution, etc. – is that 'rehabilitation' has failed to reduce crime or reform prisoners" (1981: 2).

14 See Elliott Currie (1998), *Crime and Punishment in America*.

15 The agenda of the victims' rights movement was first outlined in the 1982 "Final Report" of the President's Task Force on Victims of Crime. See also the website for the National Center for Victims of Crime: www.NCVC.org
16 See Amanda Lotz (2004), "Using 'Network' Theory."
17 See Robert Thompson (1996), *Television's Second Golden Age.*
18 While series such as *Hunter, In the Heat of the Night,* and *T.J. Hooker* were never runaway successes (often landing among the top twenty-five series, but never much higher), *CSI* is a bona fide hit, landing in the top spot in the ratings on several occasions.
19 See Stuart Kaminsky (1985), "The History and Conventions of the Police Tale"; Brooks Robards (1985), "The Police Show"; B. Keith Crew (1990), "Acting Like Cops"; James Incardi & Juliet Dee (1987), "From the *Keystone Cops* to *Miami Vice*"; John Dennington & John Tulloch (1976), "Cops, Consensus, and Ideology"; David Buxton *(1990), From The Avengers to Miami Vice*; John Sumser (1996), *Morality and Social Order in the Television Crime Drama.*
20 Julie D'Acci's analysis of *Cagney and Lacey* troubles any celebrations of the series' feminist aspirations by paying close attention to the institutional pressures that shaped the series over a number of years, changing it from a fairly traditional cop show to an exploitation-oriented "women's program." See also Steve Jenkins (1984), "Hill Street Blues"; Robert C. Allen (1985), *Speaking of Soap Operas*; Jeremy Butler (1985), "*Miami Vice*: The Legacy of Film Noir"; Danae Clark (1990), "*Cagney and Lacey*: Feminist Strategies of Detection."
21 The "Law" is here opposed to the "law": the latter representing actual laws on the books (such as laws against robbery) while the former represents the larger, social consensus – the system of morals and virtues by which we are supposed to live. Julie D'Acci discusses in excellent detail, the relationship between the protagonist and the Law:

> "The Law becomes indistinguishable from the actions and individual moral code of the cop/hero. In other words, even though the hero may break or bend institutionalized laws, the larger social order or Law is tacitly redefined to agree with the moral vision of the individual police protagonist" (1994: 115).

Importantly, this "Law" is seen as embedded in patriarchal assumptions of male power. D'Acci uses this definition as a way to underscore the masculine prerogatives of both policing and social consensus in order to engage in a larger argument about the role of female cops in crime dramas such as *Cagney and Lacey*. For D'Acci, then, the issue of the "Law" works to contain and organize the potential contradictions to male authority that are offered up by the image of a female cop.
22 See Jason Mittell (2001), "A Cultural Approach to Television Genre Theory."
23 See Rick Altman (1999), *Film/Genre*; Steve Neale (2000), *Genre and Hollywood.*
24 See Jason Mittell, "Cartoon Realism: Genre Mixing and the Cultural Life of *The Simpsons*; Jane Feuer (1992), "Genre Study and Television"; John Cawelti, "*Chinatown* and Generic transformation in Recent Films"; Thomas Schatz (1981), *Hollywood Genres: Formulas, Filmmaking, and the Studio System.*
25 Criminologist Herbert Jacob states: "When the police are called, they do not always classify the incident as criminal or make an arrest. For instance, in a neighborhood fight, the police simply may keep order and cool the situation down because it is not clear who is in the wrong or whether a crime has been committed" (1988: 50).
26 See Elayne Rapping (2003), *Law and Justice as Seen on TV.*
27 See Carol Greenhouse, Barbara Yngvesson, & David Engle, *Law and Community in Three American Towns*: "By discourse we mean an interrelated set of cultural meanings – symbols, values, and conventionalized interpretations – that shape and make comprehensible the terms people use to converse in everyday life" (1994: 2).

28 See also, Mimi White (1992), "Ideological Analysis and Television." White describes television as a "heterogeneous unity" and "regulated ideological plurality" (1992: 190–191).
29 See Amanda Lotz (2004), "Using 'Network' Theory."
30 Horace Newcomb has recently expressed his own misgivings about the "forum" model, stating that "it was unwittingly premised on a limited number of viewer options" (2000: 557). The model of the forum within this limited sphere was predicated on the "viewing strip" as the ultimate unit of analysis. Thus, in Newcomb's view the forum model fails to account for increasingly diverse and fragmented viewing practices. My own argument to the contrary is that, despite the diversity of reception, the networks themselves (both broadcast and cable) are highly invested in accurately defining and managing their potential and desired audience and, thus, program their own "strips" in an effort to keep viewers tuned into their schedule.
31 In fact, the same series moving across different networks (such as NBC to Lifetime) will likely experience a shift in the way it is interpreted. My own previous research on Lifetime's acquisition of NBC's *Homicide: Life on the Street* suggests the ways in which different discourses within the same text were given different weight according to the identity of the network on which it is viewed. The same could be said for re-runs of *Dragnet* on TV Land or "Nick at Night" which is bound to emphasize the camp elements of the series rather than its law and order seriousness. See Jonathan Nichols-Pethick (2001), "Lifetime on the Street."
32 See Anthony Cohen (1985), *The Symbolic Construction of Community*.

2 Programming the Crisis

1 John Caldwell even goes so far as to suggest that the failure of *Cop Rock* was similar in effect to that of the film *Heaven's Gate*:

> "*Cop Rock* went down in flames in the fall of 1990. Pulled from the programming wreckage, Bochco disappeared from the headlines for months following the show's collapse. Sensing a fundamental change, critics now quickly turned on the very figure who, several months earlier, could do no wrong. *Cop Rock* and Bochco evoked for television what *Heaven's Gate* and Michael Cimino had become for United Artists and the film industry almost a decade earlier: a sacrificial marquee figure, and an omen for programmers who had cut their teeth on big production values and aesthetic and financial risk" (1995: 285).

2 These terms are often used interchangeably. For my own purposes, I prefer "neo-network" and will use it to refer to this period throughout the rest of this chapter. To speak of the *post*-network era encourages (even if unintentionally) a narrative in which the major networks have finally "lost" the battle to maintain their dominance – as if they no longer exist or exert enormous pressure on the structure and function of commercial television. However, the major networks do indeed exist, and their programs continue to attract the largest percentage of the audience on a national level, though their *share* of that audience has diminished significantly. The point that I am making, while potentially a semantic argument, is that for the purposes of complicating our understanding of programming during the 1980s and 1990s (and, particularly, the renewed popularity of an older form like the police genre) we need to also complicate the "rise and fall" narrative that too often explains the "crisis" of the networks during this period. Instead of a metaphor that places the defeated networks in the past tense, we would benefit from focusing on the *neo*-network era. The neo-network era shifts emphasis from the final demise of the networks to a larger restructuring of the culture industries of which the networks are still a significant part.
3 The growth of UHF was hampered by two factors in combination. First, while the quality of the UHF image was superior to VHF, the signal itself could not be carried

as far. Thus, UHF was not as attractive to advertisers looking to reach a large audience. Second, most receivers were simply not capable of receiving a UHF signal. Many of these receivers were manufactured by RCA, which in the midst of a battle with CBS to maintain control of television's development, pushed the FCC to accept the VHF spectrum (which fitted well with RCA's technical abilities). Thus, the potential diversity that UHF promised from the outset was cut short by both political and economic determinants.

4 It is interesting to note that while a great deal has been written recently about the relationship between the major movie studios and the cable industry, very little has been written about the role of the television networks and their strategies regarding cable during this same period (roughly, 1958–1990). Of particular importance is the period from 1970 to 1987, a time when cable was experiencing its largest growth. For important studies about the rise of cable, see especially: Michele Hilmes (1990), *Hollywood and Broadcasting: From Radio to Cable*; Janet Wasko (1995), *Hollywood in the Information Age: Beyond the Silver Screen.*

5 A mid-season replacement, *Hill Street Blues* finished the season ranked 83 out of 97 prime-time network programs (Gitlin 2000: 305).

6 Webb was often referred to as a "modern day Belasco" in reference to David Belasco, the theater producer well-known for his unwavering commitment to transferring reality to the stage. See, "For *Dragnet*'s Jack Webb, Crime Pays Off." *Look Magazine.* 8 September 1953: 8–3.

7 This sequence (which concluded with the famous catchphrase, "Let's be careful out there") was lifted almost directly from Alan and Susan Raymond's 1976 PBS documentary, *The Police Tapes.* Despite the heightened style of this opening sequence, however, Steve Jenkins (1984) notes the ways in which each episode of *Hill Street Blues* rather quickly falls back into traditional Hollywood style, emphasizing the melodramatic structure of the series as a whole.

8 D'Acci covers this shift in approach in great detail in the chapter "A Woman's Program." For a detailed account of the rise of the Movie of the Week, see Elayne Rapping (1992), *The Movie of the Week: Private Stories, Public Events.*

9 *Nashville Beat* seems not to have been aired on TNN or any other station, according to Brooks and Marsh (1999).

10 For a detailed analysis of the variable "flows" of television, see Horace Newcomb (1988), "One Night of Prime Time: An Analysis of Television's Multiple Voices."

11 I am particularly influenced in this regard by studies such as Michael Curtin's *Redeeming the Wasteland* (1995), Lynn Spigel's *Make Room for TV* (1992), Christopher Anderson's *Hollywood TV* (1994), and Robert Allen's *Speaking of Soap Operas* (1985).

3 The Police Drama in Transition

1 See Newcomb & Hirsch (1983), "Television as a Cultural Forum." See also Newcomb's comments on the forum model in which he makes the case that the concept of an "arena" may be more accurate given the element of power that is largely missing from the "forum" metaphor.

2 This conservative strain is also present in films such as *Dirty Harry* and *Death Wish*, as well as television series such as *Hunter.*

3 I am not suggesting that decades are a particularly useful or even valid way to parse out television history. Gauging historical periods by decade is as arbitrary as any other temporal model. These kinds of strategies are simply heuristic devices designed to provide some element of clarity. In this case, however, the larger cycle of police programs in the 1980s and 1990s does, for the most part, break down along fairly clear lines.

4 In addition to setting the series in the same kind of dilapidated inner city that the Raymonds documented in the South Bronx of *The Police Tapes*, the makers of *Hill*

Street Blues directly lifted their signature role call sequence that began each episode from the Raymonds' film, right down to the famous catch phrase, "Let's be careful out there."

5 Class is another important variable, but it is far less diverse in this case; most cops are identified as blue-collar workers from a working-class background. It is this background that forms the basis of their "street knowledge" and it is this identification with "the street" that often causes tension when the power and duty of arrest runs up against their class position. What passes for class difference in the police drama typically occurs on two different levels: "real" felt differences between the cops and a number of the citizens that they are working to protect; and the difference in attainment between the beat cops on the street and their superiors who have worked their way up from the street, sometimes through hard work, sometimes through means that are perceived as opportunistic (i.e. taking advantage of racial or gender quotas).

6 In the introduction to a new edition of his first 87th Precinct novel, *Cop Hater* (originally published in 1956 and reissued in1999 by Pocket Books), Ed McBain (a.k.a. Evan Hunter) illustrates (and downplays) the process of generic development from his perspective:

> "But then, thinking it through further, it seemed to me that a single cop did not a series make, and it further seemed to me that something new in the annals of police procedurals (I don't even know if they were called that back then) would be a squad room *full* of cops, each with different traits, who when put together would form a *conglomerate* hero. There had been police novels before I began the 87th Precinct series. There had not, to my knowledge, been any attempt to utilize such a concept. I felt, at the time, that it was unique" (1999: xii).

For McBain, then, the process of generic transformation (with which he is often credited) comes mostly from the necessity to respond to commercial demands – the need to be unique in the market while at the same time located within a recognizable and popular form (the police novel).

7 *Miranda* rights are the result of the 1966 Supreme Court case, *Miranda v. Arizona,* which ruled that a person held for interrogation must be informed of his or her rights, such as the "right to remain silent" during interrogation and the "right to consult with a lawyer and to have the lawyer with him during interrogation" (Friedman 1993: 301–302).

8 The South Bronx that President Carter toured was the product of decades of political neglect, corruption, and short-sighted planning going back at least as far as the 1950s when 15,000 families were forced out of their homes in order to make room for the Cross-Bronx Expressway (Schur 1978: 41). By 1981, the Bronx had experienced the loss of 40 per cent of its residents (over 300,000 people) over a ten-year period due to the kinds of forcible relocation described above as well as the migration of thousands of middle-class white and minority families to the suburbs and the relocation of much of the city's manufacturing industries to other parts of the country and overseas (Purdy 1994: 47). During this period, the Bronx also experienced a rash of arson fires set by landlords hoping to collect insurance money and tenants hoping to move to the top of subsidized housing lists. And while the situation in the South Bronx was indeed "sobering," it had come to represent a larger problem: the decline of inner cities across the U.S.

9 *The Police Tapes* (73 minutes) was originally aired in 1977 on WNET in New York City and picked up later by ABC News in a slightly shorter (50 minute) version.

10 The original script for the episode called for Hill and Renko to be killed in this ambush. Test audiences and network affiliates, however, suggested that the officers survive the shooting. See Gitlin (2000); Thompson (1996); Heil (2002); Stempel (1992).

11 This scene closely echoes two key scenes in *Fort Apache, The Bronx* in which a central character – a prostitute played by Pam Grier – murders a man on the side of the road by pulling a razor from her mouth and slicing his neck. This same character also ambushed and shot two police officers earlier in the film.

12 By the mid-1980s the drug "epidemic" had begun to cross class, racial, and geographic lines (Barber 1986: 36). This spread from urban slums into middle-class suburbs prompted the Reagan administration to announce its "Campaign Against Drug Use" in August 1986, and the Congress to pass strong anti-drug legislation in September. The campaign and accompanying legislation focused mostly on domestic drug problems, establishing wide-ranging goals for awareness and prevention, especially among younger school children. "Just Say No" was the centerpiece of this part of the campaign. But the campaign and legislation also included key provisions to stem the tide of international drug trafficking.

13 The Dirty Harry films consisted of five films over a seventeen-year period: *Dirty Harry* (1971), *Magnum Force* (1973), *The Enforcer* (1976), *Sudden Impact* (1983), and *The Dead Pool* (1988). The *Death Wish* series consisted of four films over a thirteen-year span: *Death Wish* (1974), *Death Wish 2* (1982), *Death Wish 3* (1985), and *Death Wish 4: The Crackdown* (1987).

14 Most community policing initiatives emerged in the 1970s in response to rapidly increasing crime rates. According to F.B.I. statistics published in a report in *U.S. News and World Report* on July 13, 1981: total crimes in the U.S. rose by 10 per cent from 1979 to 1980, with violent crime rising 13 per cent; robbery increasing by 20 per cent, rape by 9 per cent, and murder by 7 per cent. An increase in crime was experienced in cities of all sizes, in suburbs, and in rural areas ("The People's War" 53).

15 According to Caldwell, "the stylistic emphasis that emerged during this period resulted from a number of interrelated tendencies and changes: in the industry's mode of production, in programming practices, in the audience and it's expectations, and in an economic crisis in network television" (1995: 5).

16 Though as Tom Stemple argues, while the show did borrow the look and feel of these documentaries, it did not share their narrative structure: "In both the Raymonds and Wiseman films, the pattern consists of long sequences complete in themselves, such as the door battering sequence in *The Police Tapes*, rather than *Hill Street*'s interweaving storylines, which borrow more directly from the narrative pattern of shows like *The Love Boat*" (1992: 228).

4 Stop Making Sense

1 The producers of *Homicide* borrowed this visual trope directly from David Simon's book. The Baltimore homicide units still use such a board to this day.

2 See David Simon (1997), "Murder, I Wrote."

3 For a detailed and enlightening discussion of urban renewal and gentrification, see John Teaford (2000), "Urban Renewal and Its Aftermath." Teaford highlights the lack of focus that defined Title I of the Housing Act of 1949, which was the first significant attempt at urban renewal projects. See also, Theodore Koebel (1993), "Defining, Measuring and Analyzing Community Investment." Koebel discusses urban renewal efforts after the demise of the Housing Act of 1949. In particular, he discusses Community Development Block Grants and Urban Development Action Grants which had the effect of making urban renewal an almost solely commercial project. The Inner Harbor development project in Baltimore, which resulted in such tourist attractions as Harborplace and Oriole Park at Camden Yards, was a product of these types of grants.

5 Do The Right Thing

1 See Bubeck (1995); Faulks (2000).

2 This idea was quickly dropped by Bochco before the series went into development. ABC rejected Bochco's pitch as early as February 1992: "I'm very disappointed that

we can't make the exact script we wrote and put it on the air. But I'm not naïve. I didn't actually expect that we would be able to do that" (Du Brow 1992).

3 In order to get the kind of coverage that the network desired for the series, ABC offered the pilot episode to independent and Fox affiliated station in markets where ABC's affiliates refused to carry the series. One such independent station, KTXA in Dallas, tripled its usual rating by picking up the episode (Mandese 1993: 8).

4 Despite the fact that *NYPD Blue* consistently placed in the top fifteen programs during its first season, and the network had sold all its available advertising slots, the major advertisers in the alcohol, automobile, and fast food industries were largely staying away from the show at first. The list of advertisers from the October 26th episode included: Ultra Slim Fast, Dexatrim, Alka Seltzer, Burlington Coat Factory, Plus+ White Toothpaste, Wash-N-Curl, Caruso Curls, Permathene 12, Roy Rogers Restaurants, Sharp Camcorders, and Ricola cough drops (Goldsborough 1993: 8). As ABC entertainment president Ted Harbert lamented, the series was not initially "a huge sales bonanza" (Zoglin 1985: 81).

5 Milch countered this suggestion in an interview with *Newsweek* in June of 1993: "You know how stupid that is? It's a joke. This is damned good TV. That's what we want to be judged on" (Giles & Flemming 1993: 66).

6 See Gitlin, *Inside Prime Time* (2000); Longworth, *TV Creators* (2000).

7 It is important to remember that these cable networks still relied on advertising as one of their principal sources of revenue and so tried to avoid material that would be deemed too offensive. Furthermore, at the time of writing, Congress is contemplating legislation that would place cable networks under the same content regulations as broadcasters. Interestingly, TNT is far more cautious with the series than even ABC, removing most of the mild expletives in the dialogue.

8 See Charles McGrath, "The Triumph of the Prime-Time Novel" in Newcomb (1994), *Television: The Critical View*, pp. 24–52.

9 *ER*, perhaps the single most successful hour-long drama in the history of the medium debuted the year following *NYPD Blue*'s premier. Additionally, Dick Wolf's *Law & Order*, which debuted in 1990 and has been one of the most successful hour-long dramas of the 1990s and beyond, did not crack the top-twenty in the ratings until its eighth season (1997–98).

10 Both David Caruso and Dennis Franz were nominated for Emmys during the first season (Franz won the award).

11 This analysis of the passage was offered in a sermon by Rev. Charles Haddon Spurgeon on 23 March 1856 at New Park Street Chapel, Southwark: www.spurgeon.org

12 In the first episode, Alfonse Giardella, a mobster against whom Sipowicz testified against in court, and physically assaulted in public, shot Sipowicz in a hotel room and left him for dead. This shooting occurred after Sipowicz had perjured himself with his testimony, verbally abused the Assistant District Attorney, and gone on a drinking binge; he was shot in a hotel room with a prostitute. Clearly, Andy Sipowicz began life on the series as a character in need of reconstruction and it was manifested first at the level of his body. Helen Yeates has suggested that one of the key themes of *NYPD Blue* is the vulnerability of the male body of the police detective: a figure typically associated with toughness. "In this police series, the accepted generic, patriarchal balance between being tough (extremely) and vulnerable (just a little) has been interrogated and renegotiated in favour of vulnerability and all the messiness that this word implies" (Yeates 2001: 48). Indeed, this focus on vulnerability is borne out at several points across the series. In addition to being shot, Sipowicz undergoes prostate surgery, the recovery scene of which is show in excruciating detail as Sipowicz attempts to walk for the first time after the surgery ("Prostrate Before the Law"). The surgery also leaves him impotent, a problem for which he seeks medical treatment ("Viagra Falls").

6 One Thing Leads to Another

1 The episode, "Prescription for Death," was not the original pilot episode produced for the series. That episode, "Everybody's Favorite Bagman," was aired as the fifth episode of the first season.
2 Brandon Tartikoff, President of NBC at the time of *Law & Order*'s first season once asked Dick Wolf what the "bible" of the series was. Wolf replied that it was the front page of *The New York Post* (Auster 2000: B18).
3 One notable exception to this rule, however, is *COPS*, which due to its episodic nature (as opposed to the serialized reality game shows or relationship shows) has seemingly endless syndication value.
4 So successful and valuable was *Law & Order* in its off-network run that subsequent episodes purchased by TNT (those beyond the initial 181) were purchased for roughly $700,000 per episode.
5 These were the time slots that were opened up to local affiliates by the Prime-Time Access Rule, passed in 1970.
6 Wolf spent seven years in advertising and was responsible for national campaigns for Crest toothpaste and National Airlines ("I'm Cheryl. Fly Me.").
7 These figures come from Charles Engle, Executive Vice President of Programming at Universal Media Studios, while speaking on a panel at "Law & Order: Changing Television," a symposium held at the Carsey-Wolf Center in Santa Barbara, CA on 15 April 2010.
8 The original title for the series was *Sex Crimes* and was not intended to have the *Law & Order* brand name attached to it. NBC felt that the title was too risqué and Wolf agreed to change it to the moniker used by the police force itself. Attaching it to the *Law & Order* brand was the product of a decision to capitalize on the increasing popularity of the original series as well as providing opportunities for cross-over episodes and characters.
9 Wolf makes this connection explicit: "From the very first day, we said (*Criminal Intent*) is Sherlock Holmes. That's the archetype. And that's what attracted (star) Vincent (D'Onofrio). Because Sherlock Holmes is, while very procedural, the most idiosyncratic of the major detectives. It's a voyage of discovery through his eyes. Kate (Erbe) is Dr. Watson. Jamey (Sheridan) is Lestat. It kind of breaks down very evenly" (quoted in "All 3 Shows" 02E).
10 The investigative aspects of the case are not, however, completely erased. As a way of connecting the series to the original *Law & Order*, Jerry Orbach was contracted to shift his role as Det. Lenny Briscoe to the new series. The conceit was that Briscoe had retired from the NYPD and was now working as a private investigator for the District Attorney's office.
11 Newcomb later developed this idea more fully on his own. See Newcomb (1984), "On the Dialogic Aspects of Mass Communication."
12 The phrase "especially heinous" is used during the voice-over that introduces each episode: "In the criminal justice system, sexually-based offenses are considered especially heinous. In New York City, the dedicated detectives who investigate these vicious felonies are members of an elite squad known as the Special Victims Unit. These are their stories."
13 The importance of this kind of scene to the series as a whole is explained by Rene Balcer: "At the end of an episode of *Criminal Intent*, we have a scene we call 'the aria,' which runs eight, nine, ten pages. The aria is Goren's big interrogation scene. It's a signature scene; about 75 percent of the episodes have it" (Rene Balcer quoted in Littlefield 1996: 18).
14 This point underscores the observation made first by Edgar Allen Poe, and later explained by Dennis Porter, that crime narratives are constructed in a backwards

fashion. The solution must be known before the writer can adequately lead (and mislead) the reader/writer. See Porter (1981).

15 Though it is important to recognize that the terms on which Newcomb and Hirsch constructed their argument about television as a cultural forum have changed significantly since the first publication of their article in 1983. For a clear and compelling argument about the contemporary possibilities for their theory, see Lotz (2004).

7 This Cop's For You

1 See Justin Wyatt, *High Concept*. Also, it is worth noting that perhaps the primary force in the television police drama today is producer Jerry Bruckheimer who made his name as a producer of quintessential "high concept" feature films such as *Top Gun* and *Beverly Hills Cop*.
2 See David Simon (1997), "Murder, I Wrote," *The Atlantic Monthly*.
3 *The Wire* was sold into syndication on A&E but failed to last. Other HBO series (*The Sopranos* and *Sex & The City*), however, have managed to thrive in syndication on A&E and TBS, respectively.
4 See Christopher Sterling (2004), "Communications Act of 1934," pg. 574.
5 See the FCC Fact Sheet on Program Content Regulation (http://transition.fcc.gov/mb/facts/program.html): "Section 504 of the 1996 Act requires a cable operator to fully scramble or block the audio and video portions of programming services not specifically subscribed to by a household. The cable operator must fully scramble or block the programming in question upon the request of the subscriber and at no charge to the subscriber."
6 Of course, as scholars such as Lynn Spigel, and Anna McCarthy have demonstrated, the social, cultural, and material conditions surrounding the adoption of technologies and the consumption of content have a rich and varied history of their own. See Spigel (1992), *Make Room for TV*; and McCarthy (2001), *Ambient Television*.
7 The idea of a "push" television network refers to the historic structure of programming that is made available on one primary device (the television) at a particular time. This idea is opposed to the contemporary institutional structure in which viewers are increasingly able to "pull" programming to a wide variety of viewing devices (computers, smart phones, etc.) at nearly any time.
8 See Anderson (2005), "The Uses of Drama," p. 77. See also, Nichols-Pethick (2001), "Lifetime on the Street."
9 HBO had embarked on original series programming much earlier, with series such as *The Larry Sanders Show*.
10 Ratings for the 5th season of *CSI* were up 6 per cent over the 4th season. The show's executive producer, Carol Mendelsohn, attributed that rise largely to syndication on Spike (Karrfalt 2005).
11 *CSI* is produced by CBS Paramount Television along with Alliance Atlantis and Jerry Bruckheimer Television.
12 Importantly, the audience is invited to share in this uneasiness about compromise, having been shown how compromised Mackey is as a cop, but also knowing that the suspect is guilty.
13 I need to acknowledge Vikas Kumar Gumbhir for first calling my attention to this critical possibility during his featured talk, "And All the Pieces Matter: The Reproduction of Urban Inequality in *The Wire*" presented at the First Annual International Crime, Media, and Popular Culture Studies Conference at Indiana State University, Terre Haute, IN, October 2009.
14 In fact, the episode resemble Agatha Christie's "Murder on the Orient Express." See John Cawelti (1976), *Adventure, Mystery, and Romance*.

15 See Elke Weissman & Karen Boyle (2007), "Evidence of Things Unseen" in which they make the point that "there is an implicit assumption [in the *CSI*-shot] that by seeing more, by going deeper and closer, the viewer of *CSI* is placed in a position of greater knowledge" (2007: 94).
16 See Jason Mittell (2004), *Genre and Television*.

BIBLIOGRAPHY

"ABC Goes after Five Major Market Low-Power Stations." *Broadcasting* 2 Feb. 1981: 60.

"ABC Stations Feel 'NYPD' Heat." *Christianity Today* 25 Oct. 1993: 78.

Abelman, Robert. *Reaching a Critical Mass: A Critical Analysis of Television Entertainment*. Mahwah, NJ: Lawrence Erlbaum Associates, 1998.

Albiniak, Paige. "Does That Rerun Fit the Brand?" *Broadcasting & Cable* 19 May 2008. http://www.highbeam.com/doc/1G1-179157608.html. Accessed 15 July 2011.

"All 3 Shows Have Their Own Angle." *USA Today* 6 Dec. 2002, sec. Life: 02E.

Allen, Robert C. *Speaking of Soap Operas*. Chapel Hill, NC: University of North Carolina Press, 1985.

___ ed. *Channels of Discourse, Reassembled: Television and Contemporary Criticism*. 2nd edition. Chapel Hill, NC: University of North Carolina Press, 1992.

Altman, Rick. *Film/Genre*. London: British Film Institute, 1999.

Alvarez, Raphael ed. *The Wire: Truth Be Told*. Edinburgh: Cannondale Books, 2009.

Andersen, Robin. *Consumer Culture and TV Programming*. Critical Studies in Communication and in the Cultural Industries. Boulder, CO: Westview Press, 1995.

Anderson, Christopher. *Hollywood TV: The Studio System in the Fifties*. Texas Film Studies Series. Ed. Thomas Schatz. Austin: University of Texas Press, 1994.

___. "The Uses of Drama." *Thinking Outside the Box: A Contemporary Television Genre Reader*. Eds. Gary Edgerton & Brian Rose. Lexington, KY: University Press of Kentucky, 2005. 65–90.

Auster, Albert. "A Look at Some Contemporary American and British Cops Shows." *Television Quarterly* 29.2 (1997): 46–55.

___. "Cogent Complexity Reigns on TV's Quality Tabloid." *Chronicle of Higher Education* 20 Oct. 2000: B18–19.

Bakhtin, M. M., et al. *The Bakhtin Reader: Selected Writings of Bakhtin, Medvedev, and Voloshinov*. New York: Arnold, 1994.

Balio, Tino. *Hollywood in the Age of Television*. Boston, MA: Unwin Hyman, 1990.

Barber, John. "The New Drug Crusade." *MacLean's* 29 Sept. 1986: 36–39.

Barnes, Beth E., & Lynne M. Thompson. "Power to the People (Meter): Audience Measurement Technology and Media Specialization." *Audiencemaking: How the Media Create the Audience*. Eds. James S. Ettema & D. Charles Whitney. London: Sage, 1994.

Barnouw, Erik. *Tube of Plenty*. New York: Oxford University Press, 1982.

Battaglio, Stephen. "NBC Looks to Turn on Young Adult Audience." *AdWeek* 9 Mar. 1992: 4.

Becker, Anne. "Turner Knows Branding." *Broadcasting & Cable* 23 Apr. 2007. http://www.highbeam.com/doc/1G1-162453345.html. Accessed 14 July 2011.

Benedek, Emily. "Inside *Miami Vice*: Sex and Drugs and Rock & Roll Ambush Prime-Time TV." *Rolling Stone* 28 Mar. 1985: 56–62, 125.

Blumenthal, Howard J., & Oliver R. Goodenough. *This Business of Television*. 2nd edition. New York: Billboard Books, 1998.

Booth, William. "Producer Jerry Bruckheimer: The Top Gun of Prime Time." *Washington Post* 18 Sept. 2005. http://www.washingtonpost.com/wp-dyn/content/article/2005/09/16/AR2005091600399.html. Accessed 22 Sept. 2011.

Bowman, James. "The Air Is Blue." *New Criterion* 12.2 (1993): 5–7.

Boyer, Peter J. "Production Cost Dispute Perils Hour TV Dramas." *The New York Times* 6 Mar. 1986: C26.

Brady, Shirley. "Keep the Brand in Your Hand." *Cable World*. 22 July 2002. http://www.highbeam.com/doc/1G1-89351655.html. Accessed 15 Jul. 2011.

Brennan, Steve. "Wolf Exporting *Law & Order*." *The Hollywood Reporter* 6 Apr. 1998: 7.

Brooks, Peter. *The Melodramatic Imagination: Balzac, Henry James, Melodrama, and the Mode of Excess*. New Haven, CT: Yale University Press, 1976.

Brooks, Tim, & Earl Marsh. *The Complete Directory to Prime-Time Network and Cable Shows, 1946–Present*. 7th edition. New York: Ballantine Books, 1999.

Brunsdon, Charlotte. "What is the 'Television' of Television Studies?" *The Television Studies Book*. Eds. Christine Geraghty & David Lusted. London: Arnold, 1998a. 9–14.

___. "Structure of Anxiety: Recent British Television Crime Fiction." *Screen* 39.3 (1998b): 22–3.

Bubeck, Diemut Elisabet. *Care, Gender, and Justice*. Oxford, UK: Clarendon Press, 1995.

Butler, Jeremy. "*Miami Vice*: The Legacy of Film Noir." *Journal of Popular Film and Television* 13.3 (1985): 12–8.

Buxton, David. *From the Avengers to Miami Vice: Form and Ideology in Television Series*. Manchester: Manchester University Press, 1990.

Byars, Jacki, & Eileen Meehan. "Once in a Lifetime: Constructing the 'Working Woman' Through Cable Narrowcasting." *Television: The Critical View*, 6th edition. Ed. Horace Newcomb. New York: Oxford University Press, 2000. 144–168.

Caldwell, John Thornton. *Televisuality: Style, Crisis, and Authority in American Television*. New Brunswick, NJ: Rutgers University Press, 1995.

Campbell, Christopher. "A Post-Mortem Time for Racial Imperialism." *Television Quarterly* Winter (2000): 2–0.

Carey, James W. *Communication as Culture: Essays on Media and Society*. Boston, MA: Unwin Hyman, 1989.

Carter, Bill. "Tracking Down Viewers Till They're Captured." *The New York Times* 19 Feb. 1997: C11.

___. "*Law & Order*, a Hot Franchise, Seeks a Rich Deal Early from NBC." *The New York Times* 2 Jun. 2003, sec. Business: C1.

Castro, Barry. "Middle-Management Blues: Notes from Hill Street." *Soundings* Winter 4 (1985): 43–2.

Cawelti, John G. *Adventure, Mystery, and Romance: Formula Stories as Art and Popular Culture*. Chicago, IL: University of Chicago Press, 1976.

___. "*Chinatown* and Generic Transformation in Recent American Films." *Film Theory and Criticism: Introductory Readings*, 4th edition. Eds. Gerald Mast, Marshall Cohen, and Leo Braudy. New York: Oxford University Press, 1992: 498–511.

Chris, Cynthia. "All Documentary, All the Time?: Discovery Communications, Inc. and Trends in Cable Television." *Television & New Media* 3.1 (2002): –8.

Christensen, Mark, & Cameron Stauth. *The Sweeps*. New York: William Morrow Company, 1984.

Christopher, Maurine. "Massive Cable Study Set to Roll." *Advertising Age*. 9 Feb. 1981: 2, 85.

___. "ABC Scores Cable Coup." *Advertising Age* 26 Jan. 1981: 1, 78.

___. "Why CBS Wants Cable Waiver." *Advertising Age* 2 Feb. 1981: 48.

Churchill, Bonnie. "TV's Veteran Risk Taker." *Christian Science Monitor* 2 Oct. 1990, sec. Arts: 10.

Clandfield, Peter. "'We ain't got no yard': Crime, Development, and Urban Environment." *The Wire: Urban Decay and American Television*. Eds. Tiffany Potter & C.W. Marshall. New York: Continuum, 2009. 3–9.

Clark, Danae. "*Cagney and Lacey*: Feminist Strategies of Detection." *Television and Women's Culture: The Politics of the Popular*. Ed. Mary Ellen Brown. London: Sage, 1990. 117–33.

Clarke, Alan. "'You're Nicked!': Television Police Series and the Fictional Representation of Law And Order." *Come on Down?: Popular Media Culture in Post-War Britain*. Eds. Dominic Strinati & Stephen Wagg. London: Routledge, 1992.

Cohen, Anthony P. *The Symbolic Construction of Community*. London: Tavistock, 1985.

Collins, Monica. "From the Heart; 'NYPD' Sends Smits Off in a Blaze of Raw Emotion and Great Writing." *Boston Herald* 24 Nov. 1998, sec. Arts & Life: 51.

Condit, Celeste. "The Rhetorical Limits of Polysemy." *Television: The Critical View*. Ed. Horace Newcomb. New York: Oxford University Press, 1989. 426–47.

Corliss, Richard. "Coming Up from Nowhere." *Time* 16 Sept. 1985: 64–66.

Corner, John. *Critical Ideas in Television Studies*. Oxford Television Studies. Ed. Charlotte Brunsdon & John Caughie. Oxford: Clarendon Press, 1999.

Courrier, Kevin, & Susan Green. *Law and Order: The Unofficial Companion*. Los Angeles: Renaissance Books, 1998.

Crew, B. Keith. "Acting Like Cops: The Social Reality of Crime and Law on TV Police Dramas." *Marginal Conventions: Popular Culture, Mass Media, and Social Change*. Ed. Clinton R. Sanders. Bowling Green, OH: Bowling Green State University, 1990. 131–43.

Currie, Elliott. *Crime and Punishment in America*. New York: Metropolitan Books, 1998.

Curtin, Michael. *Redeeming the Wasteland: Television Documentary and Cold War Politics. Communications, Media, and Culture*. New Brunswick, NJ: Rutgers University Press, 1995.

___. "On Edge: Culture Industries in the Neo-Network Era." *Making and Selling Culture*. Ed. Richard M. Ohmann. Hanover, NH: University Press of New England, 1996. 181–202.

___. "Feminine Desire in the Age of Satellite Television." *Journal of Communication* 49.2 Spring (1998): 5–0.

D'Acci, Julie. *Defining Women: Television and the Case of Cagney & Lacey*. Chapel Hill, NC: University of North Carolina Press, 1994.

Davis, Mike. *City of Quartz : Excavating the Future in Los Angeles*. 1st edition. New York: Vintage Books, 1992.

Day, Graham. *Community and Everyday Life*. New York: Routledge, 2006.

Delanty, Gerard. *Community*. New York: Routledge, 2003.

DeLuca, Stuart M. *Television's Transformation: The Next 25 Years*. San Diego, CA: A.S. Barnes and Company, 1980.

Deming, Caren. "*Hill Street Blues* as Narrative." *Critical Studies in Mass Communication* 2.1 (1985): 1–22.

Dempsey, John. "Advertisiers Cop Out on FX's Edgy Shield." *Daily Variety* 10 Apr. 2002. http://www.highbeam.com/doc/1G1-84876284.html. Accessed 22 Sept. 2011.

Denisoff, R. Serge. *Inside MTV*. New Brunswick, NJ: Rutgers University Press, 1990.

Denning, Michael. *Mechanic Accents: Dime Novels and Working-Class Culture in America. The Haymarket Series*. Revised edition. London; New York: Verso, 1998.

Dennington, John, & John Tulloch. "Cops, Consensus, and Ideology." *Screen Education* 20 Autumn (1976): 37–47.

Dolan, Marc. "The Peaks and Valleys of Serial Creativity: What Happened to/on *Twin Peaks*." *Trust No One: Critical Approaches to Twin Peaks*. Ed. David Lavery. Detroit, MI: Wayne State University Press, 1995. 30–50.

Dove, George N. *The Police Procedural*. Bowling Green, OH: Bowling Green University Popular Press, 1982.

Du Brow, Rick. "Can the Bochco-ABC Marriage Be Saved?" *Los Angeles Times*. 15 Feb, 1992. http://articles.latimes.com/1992-02-15/entertainment/ca-1604_1_abc-marriage. Accessed 11 Dec. 2011.

Dudsic, Greg. "NBC Orders Up Spinoff Series." *Daily Variety* 5 May 1999: A4.

Duffy, James. *Stay Tuned*. New York: Dunhill Publishing Co., 1997.

Edgerton, Gary R. *Television Histories: Shaping Collective Memory in the Media Age*. Lexington, KY: University Press of Kentucky, 2001.

Elliott, Stuart. "Controversy May Sell, but Only a Few Marketers Took Chance with 'NYPD Blue.'" *The New York Times* 23 Sept. 1993: D19.

Ely, Melvin Patrick. *The Adventures of Amos 'N' Andy : A Social History of an American Phenomenon*. New York: Free Press, 1991.

Faulks, Keith. *Citizenship*. London: Routledge, 2000.

"FCC Is Asked to Take Warner out of Pay Cable." *Broadcasting* 25.1 (1974): 31.

Feuer, Jane. "Melodrama, Serial Form, and Television Today." *Screen* 25.1 (1984): 4–16.

___. "Genre Study and Television." *Channels of Discourse, Reassembled*. Ed. Robert C. Allen. Chapel Hill, NC: University of North Carolina Press, 1992. 138–60.

___. *Seeing Through the Eighties: Television and Reaganism*. Durham, NC: Duke University Press, 1995.

Fishman, Jessica. "The Populace and the Police: Models of Social Control in Reality-Based Crime Television." *Critical Studies in Mass Communication* 16 (1999): 268–88.

Fiske, John. *Television Culture*. London: Routledge, 1987.

___. *Media Matters: Everyday Culture and Political Change*. Minneapolis, MN: University of Minnesota Press, 1994.

Fiske, John, & John Hartley. *Reading Television*. London: Methuen, 1978.

Fletcher, Connie. *What Cops Know*. New York: Pocket Books, 1990.

Fontana, Tom. Interview. *NBC.COM: Homicide: Life on the Set* 7 Feb. 1998. http://www.nbc.com/homicide/set/fontana/hm_text02b.html

"For *Dragnet*'s Jack Webb Crime Pays Off." *Look* 8 Sept. 1953: 88–93.

Freeman, Mike. "Hour Drama Boldly Going to First-Run." *Broadcasting* 17 Feb. 1992: 22–24.

Foucault, Michel. *Discipline and Punish: The Birth of the Prison*. 2nd edition. New York: Vintage Books, 1995.

Fried, Joseph P. "The South Bronx, U.S.A." *The New York Times* 7 Oct. 1977: 27.

Friedman, Lawrence M. *Crime and Punishment in American History*. New York: Basic Books, 1993.

Garnham, Nicholas. *Capitalism and Communication: Global Culture and the Economics of Information*. The Media, Culture & Society Series. Ed. John Corner, et al. London: Sage Publications, 1990.

Garofalo, James, & Maureen McLeod. "The Structure and Operations of Neighborhood Watch Programs in the United States." *Crime and Delinquency* 35.3 (1989): 326–44.

Garvin, Glenn. "'Shield' is TV's Hottest – and Most Controversial – Cop Show." *Knight Ridder/Tribune News Service* 13 May 2002. http://www.highbeam.com/doc/1G1-86240143.html. Accessed 22 Sept. 2011.

Gilbert, Matthew. "Those True to 'Blue' Will Like the Opener." *Boston Globe* 30 Sept. 1997, sec. Living: E1.

Giles, Dennis. "A Structural Analysis of the Police Story." *American Television Genres*. Eds. Stuart Kaminsky & Jeffrey H. Mahan. Chicago: Nelson-Hall, 1985. 67–84.

Giles, Jeff & Charles Flemming. "The Wild Men of Prime Time." *Newsweek* 28 Jun., 1993: 66.

Giltenan, E. "Tarnished Silverman." *Forbes* 30 Mar. 1992: 14.

Gitlin, Todd. "Prime-Time Ideology: The Hegemonic Process in Television Entertainment." *Television: The Critical View*. Ed. Horace Newcomb. New York: Oxford University Press, 1979. 516–36.

___. *Watching Television: A Pantheon Guide to Popular Culture*. 1st ed. New York: Pantheon Books, 1987.

___. *Inside Prime Time*. 4th edition. Berkeley, CA: University of California Press, 2000.

Glassner, Barry. *The Culture of Fear: Why Americans Are Afraid of the Wrong Things*. New York: Basic Books, 1999.

Gledhill, Christine. "Speculations on the Relationship between Soap Opera and Melodrama." *Quarterly Review of Film and Video* 14.– (1992): 103–24.

Goldenson, Leonard, & Marvin J. Wolf. *Beating the Odds: The Untold Story Behind the Rise of ABC*. New York: Charles Scribner's Sons, 1991.

Goldsborough, Robert. "Beavis Pales Next to TV News." *Advertising Age*. 1 Nov. 1993: 8.

Goodman, Tim. "When TV Brands Go Off Brand." *The Hollywood Reporter* 24 Nov. 2010. http://www.hollywoodreporter.com/news/tv-brands-brand-47791. Accessed 12 July 2011.

Gray, Herman. *Watching Race: Television and the Struggle for "Blackness."* Minneapolis, MN: University of Minnesota Press, 1995.

Greenhouse, Carol, Barbara Yngvesson, & David M. Engel. *Law and Community in Three American Towns*. Ithica, NY: Cornell University Press, 1994.

Grossberg, Lawrence. "Cultural Studies and/in New Worlds." *Critical Studies in Mass Communication* 10.1 (1993): 1–22.

Gubernick, Lisa. "I'm Not Rumpled Anymore." *Forbes* 6 Mar. 1989: 84–85.

Guider, Elizabeth. "'Brody,' 'Columbo' Wow Int'l Buyers; 'Law' a Harder Sell." *Daily Variety* 23 May 2001: 7.

Gumbhir, Vikas Kumar. "And All the Pieces Matter: The Reproduction of Urban Inequality in *The Wire*." First Annual International Crime, Media, and Popular Culture Studies Conference. Indiana State University, Terre Haute, IN, October 2009.

Hall, Stuart. "The Problem of Ideology: Marxism without Guarantees." *Stuart Hall: Critical Dialogues in Cultural Studies*. Eds. David Morley & Kuan-Hsing Chen. London: Routledge, 2003. 25–46.

Halttunen, Karen. *Murder Most Foul: The Killer and the American Gothic Imagination*. Cambridge, MA: Harvard University Press, 1998.

Hampp, Andrew. "FX Breaks Out of the Box." *AdAge MediaWorks* 11 Dec. 2007. http://adage.com/article/mediaworks/fx-breaks-box/122539/. Accessed 14 Jul. 2011.

___. "How USA Network Built 'Character.'" *Advertising Age* 17 May 2010. http://www.highbeam.com/doc/1G1-226798802.html. Accessed 18 Sept. 2011.

Hanczor, Robert. "Articulation Theory and Public Controversy: Taking Sides over *NYPD Blue*." *Critical Studies in Mass Communication* 14.1 (1997): 1–30.

Hanson, Janice. "The View from the Hill: *Hill Street Blues*." *Clues* Spring/Summer (1984): 58–72.

Harvey, David. *The Condition of Postmodernity: An Enquiry into the Origins of Cultural Change*. Oxford, England; New York, NY: Blackwell, 1989.

Heil, Douglas. *Prime-Time Authorship*. Syracuse, NY: Syracuse University Press, 2002.

Hettrick, Scott. "This Fall: 'Law & Order' on A&E." *The Hollywood Reporter* 27 Apr. 1994. http://web.lexis-nexis.com/universe/document?_m=2f23cd5428a8c5753ac4b-13d783c362c&_documen=32&wchp=dGLbVtz-zSkVA&_md5=03b0d8f63dd. Accessed 9 June 2005.

Hill, Michael. "Milner & McCord: A Family Cop Show with a Twang." *The Washington Post* 22 Oct. 1989: 45.

Hilmes, Michele. *Hollywood and Broadcasting: From Radio to Cable*. Urbana, IL: University of Illinois Press, 1990.

Hoffman, Tod. *Homicide: Life on the Screen*. Toronto: ECW Press, 1998.

Horton, John. "The Rise of the Right: A Global View." *Crime and Social Justice* 15.1 (1981): 7–17.

"How Cable-TV Success Hinges on Satellites." *Business Week* 14 Sept. 1981: 89–90.

Huntington, Samuel. *American Politics: The Promise of Disharmony*. Cambridge, MA: Harvard University Press, 1981.

___. "The Clash of Civilizations?" *Foreign Affairs* 6.3 (1993): 22–49.

Hurd, Geoffrey. "The Television Presentation of the Police." *Popular Television and Film*. Ed. Tony Bennett. London: BFI, 1981. 53–70.

Incardi, James A., & Juliet L. Dee. "From the Keystone Cops to *Miami Vice*: Images of Policing in American Popular Culture." *Journal of Popular Culture Fall* (1987): 84–102.

Ivey, Mark. "Suddenly, Basic Cable Is Offering More Than Basic Fare." *Business Week* 7 Mar. 1988: 36.

Jacob, Herbert. *Crime and Justice in Urban America*. Englewood Cliffs, NJ: Prentice-Hall, 1980.

Jaramillo, Deborah L. "The Family Racket: AOL, Time Warner, HBO, *The Sopranos*, and the Construction of a Quality Brand." *Journal of Communication Inquiry* 26.1 (2002): 59–75.

Jenkins, Steve. "*Hill Street Blues*." *MTM: Quality Television*. Eds. Jane Feuer, Paul Kerr & Tise Vahimagi. London: BFI Publishing, 1984. 183–99.

Jermyn, Deborah, "Body Matters: Realism, Spectacle and the Corpse in *CSI*." *CSI: Crime TV Under the Microscope*. Ed. Michael Allen. New York: I.B. Tauris, 2007. 79–89.

Johnson, Peter. "Is ABC Ready to Pay for 26 More 'Peaks'?" *USA Today* 24 Apr. 1990, sec. Life: 3D.

Jones, Dylan. "Nielsens: ABC Claims Rare Sunday Win." *USA Today* 11 Apr. 1990, sec. Life: 3D.

Jubera, Drew. "Post-Hiatus 'NYPD Blue' Gets Its Gritty Edge Back." *Atlanta Journal and Constitution* 11 Jan. 2000: 1C.

Kaminer, Wendy. *It's All the Rage: Crime and Culture.* Reading, MA: Addison-Wesley Publishing Co., 1995.

Kaminsky, Stuart. "The History and Conventions of the Police Tale." *American Television Genres.* Eds. Stuart Kaminsky & Jeffrey H. Mahan. Chicago: Nelson-Hall, 1985. 53–66.

Kaplan, E. Ann. *Rocking Around the Clock: Music Television, Postmodernism, and Consumer Culture.* New York: Routledge, 1987.

Karrfalt, Wayne. "CSI: A 'Manly' Pursuit." *Television Week* 30 May 2005. http://www.highbeam.com/doc/1G1-132978342.html. Accessed 22 Sept. 2011.

Katz, Jon. "Covering the Cops." *Columbia Journalism Review* January/February 1993: 25–30.

Katz, Richard. "TNT Snags 'Law' Rights." *Daily Variety* 7 Jan. 1999: 1.

Keetley, Dawn. "Law & Order." *Prime Time Law: Fictional Television as Legal Narrative.* Eds. Robert M. Jarvis & Paul R. Joseph. Durham, NC: Carolina Academic Press, 1998. 33–53.

Kerr, Paul. "Drama at MTM: *Lou Grant* and *Hill Street Blues.*" *MTM: Quality Television.* Eds. Jane Feuer, Paul Kerr & Tise Vahimagi. London: BFI, 1984.

Kiesel, Diane. "Crime and Punishment: Victims' Rights Movement Presses Courts, Legislatures." *American Bar Association Journal* 70.1 (1984): 25–28.

Koebel, Theodore. *Defining, Measuring and Analysing Community Investment.* Center For Housing Research, 1993.

Kompare, Derek. *CSI.* Malden, MA: Wiley-Blackwell, 2010. Kindle Edition.

Landrum, Larry. "Instrumental Texts and Stereotyping in *Hill Street Blues*: The Police Procedural on Television." *Melus* 11.3 (1984): 93–100.

Lawrence, Regina. *The Politics of Force: Media and the Construction of Police Brutality.* Berkeley: University of California Press, 2000.

Leland, John, & Charles Fleming. "Blue in the Night." *Newsweek* 13 Dec. 1993: 56–59.

Levin, Gary. "Plot Ideas Ripped From the Headlines." *USA Today* 5 Dec. 2002: 01E.

Levinson, Barry. Interview. *Guardian Unlimited.* 7 Sept. 2000. http://film.guardian.co.uk/interview/interviewpages/0,6737,367281,00.html. Accessed 20 Nov. 2004.

Lindheim, Richard D., and Richard A. Blum. *Inside Television Producing.* Boston: Focal Press, 1991.

Lister, Ruth. *Citizenship: Feminist Perspectives,* 2nd edition. New York: New York University Press, 2003.

Littlefield, Kinney. "'Law & Order'; the Blue-Chip Drama Goes Strong into Its Seventh Season." *Austin American-Statesman* 6 Oct. 1996, sec. Entertainment: 5.

___. "Behind the Scenes of *Law & Order.*" *Writer* 2005: 18–22.

Littleton, Cynthia. "Producers: More TV Dramas to Come." *United Press International.* 21 Mar. 1995. http://web.lexis-nexis.com/universe/document? m=175c788687673cbf845888abdac44b75& docum=12&wchp=dGLbVlb-zSkVA& md5=6c9e4c13057998abb594d43d1dff34a0. Accessed 31 May 2005

___. "Cable Targets Global 'Niche' Brand." *Broadcasting & Cable* 1 July 1996. http://www.highbeam.com/doc/1G1-18442585.html. Accessed 15 Jul. 2011.

___. "Wolf at Intl Door for *Law.*" *Daily Variety* 21 Jan. 1998: 6.

Longworth, James L. *TV Creators: Conversations with America's Top Producers of Television Drama.* Syracuse, NY: Syracuse University Press, 2000.

Lotz, Amanda D. "Using 'Network' Theory in the Post-Network Era: Fictional 9/11 U.S. Television Discourse as a Cultural Forum." *Screen* 45.4 (2004): 423–39.

___. *Redesigning Women: Television After the Network Era.* Urbana, IL: University of Illinois Press, 2006.

___. "If It's Not Television, What Is It?" *Cable Visions: Television Beyond Broadcasting.* Eds. Sarah Banet-Weiser, Cynthia Chris, & Anthony Freitas. New York: New York University Press, 2007a. 112–132.

___. *The Television Will Be Revolutionized.* New York: New York University Press, 2007b.

MacDonald, J. Fred. *One Nation Under Television: The Rise and Decline of Network TV.* New York: Pantheon Books, 1990.

Mahler, Richard. "2 CBS Shows Seek to Drop Serial Formats." *Electronic Media* 1 Feb. 1988: 83.

Mahoney, William. "Fox Picks Up 'Cops' Series for Saturday." *Electronic Media* 6 Mar. 1989: 16.

Mahoney, William, & Richard Mahler. "'St. Elsewhere' Autopsy: Did a Bleak Syndication Future Kill Show?" *Electronic Media* 25 Jan. 1988: 3.

Mair, George. *Inside HBO: The Billion Dollar War between HBO, Hollywood, and the Home.* New York: Dodd, Mead and Co., 1988.

Malone, Patrick. "'You Have the Right to Remain Silent': Miranda After Twenty Years." *The American Scholar* 55.3 (1986): 367–380.

Mandese, Joe. "Independent Finds 'NYPD' Arresting." *Advertising Age* 1 Nov. 1993: 8.

___. "'NYPD Blue' Showdown." *Advertising Age* 21 Mar. 1994: 1, 42.

Marc, David. *Demographic Vistas.* Revised Edition. Philadelphia, PA: University of Pennsylvania Press, 1996.

Martin, Julie. Personal Interview. 5 Feb. 2010.

Martinez, Elizabeth, & Arnoldo Garcia. "What is Neoliberalism?" 1 Jan. 1997. http://www.corpwatch.org/article.php?id=376. Accessed 17 Jan. 2005.

Marvin, Carolyn. *When Old Technologies Were New: Thinking About Electric Communication in the Late Nineteenth Century.* New York: Oxford University Press, 1988.

Massey, Doreen. "A Place Called Home?" *New Formations.* Summer (1992): 3–15.

McAllister, Matthew. "Financial Interest and Syndication Rules." *Museum of Broadcast Communications Encyclopedia of Television,* 2nd edition. Ed. Horace Newcomb. New York: Fitzroy-Dearborn, 1997: 875–877.

McBain, Ed. *Cop Hater.* New York: Pocket Books, 1999.

McCarthy, Anna. *Ambient Television: Visual Culture and Public Space.* Durham, NC: Duke University Press, 2001.

McConnell, Frank. "Smart, Hip and Real: Bochco's 'NYPD Blue.'" *Commonweal* 8 Oct. 1993: 20–21.

McConville, Jim. "In Branding, Image is Reality." *Broadcast & Cable* 4 Dec. 1995. http://www.highbeam.com/doc/1G1-17842962.html. Accessed 12 Jul. 2011.

Meehan, Eileen. "Conceptualizing Culture as a Commodity." *Television: The Critical View,* 5th edition. Ed. Horace Newcomb. New York: Oxford University Press, 1986. 563–72.

Meehan, Eileen, & Jackie Byars. "Telefeminism: How Lifetime Got Its Groove, 1984–1997." *Television & New Media* 1.1 (2000): 33–51.

Meisler, Andy. "A New Face at 'Law & Order.' Any Objections?" *The New York Times* 25 Sept. 1994, sec. Art and Leisure: 33.
"MGM Domestic Television Distribution's *In the Heat of the Night* Premiers in Syndication Monday, September 21." *PR Newswire* 9 Sept. 1992.
Milch, David, & Bill Clark. *True Blue: The Real Stories Behind NYPD Blue*. New York: Willliam Morrow and Company, Inc., 1995.
Mills, C. Wright. *The Sociological Imagination*. New York: Oxford University Press, 1959.
Mittell, Jason. "Cartoon Realism: Genre Mixing and the Cultural Life of *The Simpsons*." *The Velvet Light Trap* 47 (2001a): 1–8.
___. "A Cultural Approach to Television Genre Theory." *Cinema Journal* 40.3 (2001b): –4.
___. *Genre and Television: From Cop Shows to Cartoons in American Culture*. New York: Routledge, 2004.
Moore, Frazier. "FX Targets Male Viewers." *AP Online* 14 June 1999. http://www.highbeam.com/doc/1P1-23229254.html. Accessed 12 Jul. 2011.
Morley, David, & Kevin Robins. *Spaces of Identity: Global Media, Electronic Landscapes, and Cultural Boundaries*. London: Routledge, 1995.
Mosco, Vincent. *The Political Economy of Communication: Rethinking and Renewal*. London: Sage Publications, 1996.
Munby, Jonathan. *Public Enemies, Public Heroes: Screening the Gangster from Little Caesar to Touch of Evil*. Chicago, IL: University of Chicago Press, 1999.
MTV Music Television. Advertisement. *Television/Radio Age* 20 Sept. 1982a: 5. Print.
MTV Music Television. Advertisement. *Television/Radio Age* 15 Nov. 1982b: 13. Print.
Naremore, James. *More Than Night: Film Noir in Its Contexts*. Berkeley, CA: University of California Press, 1998.
"NBC, 1981–85: The Climb to the Top." *Broadcasting* 9 Jun. 1986: 80–83.
Neale, Steve. *Genre and Hollywood*. London: Routledge, 2000.
New Television Networks: Entry, Jurisdiction, Ownership, and Regulation. Washington, DC: Federal Communications Commission, 1980.
Newcomb, Horace. *Television: The Most Popular Art*. Garden City, NJ: Anchor Press, 1974.
___. "On the Dialogic Aspects of Mass Communication." *Critical Studies in Mass Communication* 1.1 (1984): 34–50.
___. "One Night of Prime Time: An Analysis of Television's Multiple Voices." *Media, Myths, and Narratives: Television and the Press*. Ed. James W. Carey. *Sage Annual Reviews of Comunication Research*. Newbury Park, CA: Sage Publications, Inc., 1988. 88–112.
___, ed. *Television: The Critical View*. 5th edition. New York: Oxford University Press, 1994.
___, ed. *Television: The Critical View*. 6th edition. New York: Oxford University Press, 2000.
Newcomb, Horace, & Paul M. Hirsch. "Television as a Cultural Forum." *Television: The Critical View*. Ed. Horace Newcomb. New York: Oxford University Press, 1983. 503–15.
Nichols, Bill. *Ideology and the Image: Social Representation in the Cinema and Other Media*. Bloomington, IN: Indiana University Press, 1981.
___. *Representing Reality*. Bloomington, IN: Indiana University Press, 1991.
Nichols-Pethick, Jonathan. "Lifetime on the Street: Textual Strategies of Syndication." *The Velvet Light Trap* 47 (2001): 62–73.

___. "Homicide: Life on the Street." *Museum of Broadcast Communications Encyclopedia of Television*, 2nd edition. Ed. Horace Newcomb. New York: Fitzroy Dearborn, 2004: 112–125.

Oates, Joyce Carol. "For Its Audacity, Its Defiantly Bad Taste, and Its Superb Character Studies." *TV Guide* 1 Jun. 1985: 4–7.

Ouellette, Laurie. "'Take Responsibility for Yourself': *Judge Judy* and the Neoliberal Citizen." *Reality TV: Remaking Television Culture*. Eds. Susan Murray & Laurie Ouellette. New York: New York University Press, 2004. 231–50.

Owen, Rob. "Gritty Drama *The Shield* Brings FX Network's Niche Into Focus." *Pittsburgh Post-Gazette*. 5 Jan. 2003.

___. "Cable Networks Brand Themselves Through the Look and Feel of Programs." *Pittsburgh Post-Gazette*. 24 Jul. 2011. http://www.post- gazette.com/pg/11205/1162020-67-0.stm. Accessed 25 July 2011.

Papazian, Ed, ed. *TV Dimensions '89*. New York: Media Dynamics, Inc., 1989.

Parry-Giles, Trevor, & Paul J. Traudt. "The Depiction of Television Courtroom Drama: A Dialogic Criticism of *L. A. Law*." *Television Criticism*. Eds. Leah R. VandeBerg and Lawrence Wenner. White Plains, NY: Longman, 1991. 143–159.

"The People's War against Crime." *U.S. News and World Report* 13 Jul. 1981: 53.

Pierce, Scott D. "Raising 'The Shield.'" *Deseret News* 29 Aug. 2008. http://www.highbeam.com/doc/1P2-17095267.html. Accessed 22 Sept. 2011.

Plasketes, George. "*Cop Rock* Revisted: Unsung Series and Musical Hinge in Cross-Genre Evolution." *Journal of Popular Film and Television* 32.2 (2004): 64–73.

Platt, Tony, & Paul Takagi. "Law and Order in the 1980s." *Crime and Social Justice* 15.1 (1981): 1–6.

Plummer, Ken. *Intimate Citizenship: Private Decisions and Public Dialogues*. Seattle, WA: University of Washington Press, 2003.

Pollan, Michael. "Can *Hill Street Blues* Rescue NBC?" *Channels* Mar./Apr. (1983): 30–34.

Poniewozik, James. "New Cops on the Beat." *Time* 10 Jun. 2002: 58–59.

Porter, Dennis. *The Pursuit of Crime: Art and Ideology in Detective Fiction*. New Haven, CT: Yale University Press, 1981.

Potter, Tiffany & C.W. Marshall, eds. *The Wire: Urban Decay and American Television*. New York: Continuum, 2009.

Pribram, Deidre. "Viewer Discretion Advised: Moral and Emotional Codes in *NYPD Blue*." *Creative Screenwriting* Dec. 1997: 104–18.

"Promises for Pay-TV Become Reality in New Orleans." *CATV* 28 Apr. 1975: 7–8.

"PTC Attacks 'Shield.'" *Broadcasting & Cable* 10 Apr. 2002. http://www.highbeam.com/doc/1G1-85045058.html. Accessed 22 Sept. 2011

Purdy, Matthew. "Left to Die, the South Bronx Rises from Decades of Decay." *The New York Times* 13 Nov. 1994: 1, 47.

Quinn, Laura. "The Politics of *Law and Order*." *The Journal of American & Comparative Cultures* 25.– (2002): 130–133.

Rapping, Elayne. *The Movie of the Week: Private Stories, Public Events*. Minneapolis, MN: University of Minnesota Press, 1992.

___. "Cops, Crime, and TV." *Progressive* Apr. 1994: 36–38.

___. *Law and Justice as Seen on TV*. New York: New York University Press, 2003.

Reagan, Ronald. *Remarks Announcing the Campaign Against Drug Abuse*. 1986. University of Texas. Available: www.reagan.utexas.edu/archives/speeches/1986/080486b.htm. 14 Oct. 2005.

Reiss, Albert J., Jr. *The Police and the Public*. 1971. 6th edition. New Haven: Yale University Press, 1977.

Research and Forecast, Inc., & Ardy Friedberg. *America Afraid: How Fear of Crime Changes the Way We Live*. New York: New American Library, 1983.

Richmond, Ray. "Cable's Net Worth." *Variety* 21 Jan. 1998: 108.

Robards, Brooks. "The Police Show." *TV Genres: A Handbook and Reference Guide*. Ed. Brian Geoffrey Rose. Westport, CT: Greenwood Press, 1985. 11–25.

Rohrer, Trish. "Escape from L.A." *New York* 14 Jul. 1997: 38–43.

Ross, Ken. "Producing Excellence (an Interview with Dick Wolf)." *Produced By* Fall 2000: 1–2.

Rothenberg, Randall. "CBS Tries 'Branding' Its Shows." *New York Times* 10 Nov. 1988: D22

Schatz, Thomas. *Hollywood Genres: Formulas, Filmmaking, and the Studio System*. Philadelphia: Temple University Press, 1981.

___. "*St. Elsewhere* and the Evolution of the Ensemble Series." *Television: The Critical View*. Ed. Horace Newcomb. 4th edition. New York: Oxford University Press, 1987. 85–100.

Schiff, Laura. "Maestro in Blue." *Creative Screenwriting* Dec. 1997: 3–10.

Schur, Robert. "Prescription for the South Bronx." *The Nation* 21 Jan. 1978: 38–42.

Seagal, D. Letter. *Columbia Journalism Review* Mar./Apr. 1993: 4–5.

Selznick, Barbara. "Branding the Future: SyFy in the Post-Network Era." *Science Fiction Film and Television*. 2:2 (2009): 177–204.

Shales, Tom. "The Re-Decade." *Esquire* March 1986: 67–72.

Sherman, Jay. "Cable Debates Originals vs. Off-Net." *Television Week* 1 Sept., 2003. http://www.highbeam.com/doc/1G1-107311668.html. Accessed 15 Jul. 2011.

Simon, David. *Homicide: A Year on the Killing Streets*. New York: Ivy Books, 1991.

___. "The Reporter I: Cops, Killers, and Crispy Critters." *The Culture of Crime*. Eds. Craig L. LaMay & Everette E. Dennis. New Brunswick, NJ: Transaction Publishers, 1995. 35–45.

___. "Murder, I Wrote." *The New Republic* 8 Sept. 1997: 18–20.

___. "Introduction" and "Letter to HBO." *The Wire: Truth Be Told*. Ed. Raphael Alvarez. Edinburgh: Cannondale Books, 2009. 1–36.

Slater, Robert. *This Is...CBS: A Chronicle of 60 Years*. Engelwood Cliffs, NJ: Prentice-Hall, 1988.

Smith, Brent L., John J. Sloan, & Richard M. Ward. "Public Support for the Victims' Rights Movement: Results of a Statewide Survey." *Crime and Delinquency* 36.4 (1990).

Smith, Sally Bedell. "*Miami Vice*: Action TV with Some New Twists." *The New York Times* 3 Jan. 1985: C20.

Sparks, Richard. *Television and the Drama of Crime: Moral Tales and the Place of Crime in Public Life. New Directions in Criminology Series*. Philadelphia, PA: Open University Press, 1992.

Spigel, Lynn. *Make Room for TV: Television and the Family Ideal in Postwar America*. Chicago: University of Chicago Press, 1992.

___. *The Revolution Wasn't Televised: Sixties Television and Social Conflict*. New York: Routledge, 1997.

Spillman, Susan. "USA 'Gongs' Original Shows." *Advertising Age* 17 Sept. 1984: 87.

Stanley, T.I. "Cutting-Edge FX Hailed as Next HBO." *Advertising Age* 11 Oct. 2004. http://adage.com/article/news/cutting-edge-fx-hailed-hbo/100760/. Accessed 14 Jul. 2011.

Stempel, Tom. *Storytellers to the Nation: A History of American Television Writing.* New York: Continuum, 1992.

Sterling, Christopher. "Communications Act of 1934." *Museum of Broadcast Communications Encyclopedia of Television*, 2nd edition. Ed. Horace Newcomb. New York: Fitzroy Dearborn, 2004: 574–575.

Sterling, Christopher, & John M. Kitross. *Stay Tuned: A Concise History of American Broadcasting.* 2nd edition. Belmont, CA: Wadsworth Publishing, 1990.

Steward, Tom. "Making the Commercial Personal: The Authorial Value of Jerry Bruckheimer Television." *Continuum: Journal of Media & Cultural Studies.* 24:5 (2010): 735–749.

Streeter, Thomas. "The Cable Fable Revisited: Discourse, Policy, and the Making of Cable Television." *Critical Studies in Mass Communication* 4 (1987): 174–200.

___. *Selling the Air: A Critique of the Policy of Commercial Broadcasting in the United States.* Chicago, IL: University of Chicago Press, 1996.

Sumser, John. *Morality and Social Order in Television Crime Drama.* Jefferson, NC: McFarland, 1996.

Svetkey, Benjamin. "The Long Arm of the Law." *Entertainment Weekly* 12 Nov. 1999: 26–33.

Tartikoff, Brandon, & Charles Leerhsen. *The Last Great Ride.* New York: Turtle Bay Books, 1992.

Taylor, Clarke. "TV Generation Key to Cable Success?" *The Los Angeles Times* 18 Apr. 1981: B8.

Teaford, Jon. "Urban Renewal and Its Aftermath." *Housing Policy Debate* 11.2 (2000). 443–465.

Thompson, Robert J. *Adventures on Prime Time.* Media and Society. Ed. J. Fred MacDonald. New York: Praeger, 1990.

___. *Television's Second Golden Age: From Hill Street Blues to E.R.* New York: Continuum, 1996.

Thomson, David. "Rough Love." *Esquire* Nov. 1998: 64–66.

"Three Views from the Top." *Broadcasting* 83 (1974): 21–23.

Tinker, Grant, & Bud Rukeyser. *Tinker in Television: From General Sarnoff to General Electric.* New York: Simon & Schuster, 1994.

Turow, Joseph. *Breaking up America: Advertisers and the New Media World.* Chicago, IL: University of Chicago Press, 1997.

Tyrer, Thomas. "'Order' Thrives Despite Changes." *Electronic Media* 28 Feb. 1994: 40.

Unger, Irwin & Debbie Unger. *Twentieth Century America.* New York: St. Martin's Press, 1990.

Uviller, H. Richard. "The View from a Baltimore Squad Car." *The Washington Post* 9 Jun. 1991, sec. Book World: X4.

Walley, Wayne. "Cop Dramas Bow to Real-Life Crime." *Advertising Age* 10 Apr. 1989: 42.

Walley, Wayne, & Thomas Tyrer. "'Order' Skips Syndication for A&E." *Electronic Media* 17 Jan. 1994: 3.

Wasko, Janet. *Hollywood in the Information Age: Beyond the Silver Screen. Texas Film Studies Series.* 1st University of Texas Press edition. Austin, TX: University of Texas Press, 1995.

Waters, Harry. "*Miami Vice*: Pop and Cop." *Newsweek* 21 Jan. 1985: 67.

___. "Return of a Network Reject." *Newsweek* 6 Jun. 1988: 60–61.

Webster, James, Patricia Phalen, & Lawrence Lichty. *Ratings Analysis: The Theory and Practice of Audience Research*. 2nd edition. Mahwah, NJ: Lawrence Erlbaum Associates, 2000.

Weisman, Jon. "Tube Testosterone: Cabler Lures Highly Sought Male Demo." *Variety* 8 Jun. 2004. http://www.highbeam.com/doc/1G1-118687049.html. Accessed 12 Jul. 2011.

Weissmann, Elke & Karen Boyle, "Evidence of Things Unseen: The Pornographic Aesthetic and the Search for Truth in *CSI*." *CSI: Crime TV Under the Microscope*. Ed. Michael Allen. New York: I.B. Tauris, 2007. 90–102.

White, Mimi. "Ideological Analysis and Television." *Channels of Discourse, Reassembled: Television and Contemporary Criticism*. Ed. Robert C. Allen. 2nd edition. Chapel Hill, NC: University of North Carolina Press, 1992. 161–202.

"Wildmon Targets New Cop Show." *Christianity Today* 16 Aug. 1993: 42.

Williams, Linda. "Melodrama Revised." *Refiguring American Film Genres: Theory and History*. Ed. Nick Browne. Berkeley, CA: University of California Press, 1998. 42–88.

Williams, Raymond. *Television: Technology and Cultural Form*. Hanover, NH: Wesleyan University Press, 1992.

Wilson, Christopher P. *Cop Knowledge: Police Power and Cultural Narrative in Twentieth Century America*. Chicago, IL: University of Chicago Press, 2000.

Wilson, James Q., & George L. Kelling. "Broken Windows: The Police and Neighborhood Safety." *Atlantic Monthly* Mar. 1982: 29–38.

Wolcott, James. "Untrue Grit." *New Yorker* 4 Oct. 1993: 217–19.

Wolf, Dick. *Law & Order Crime Scenes*. New York: Barnes & Noble Books, 2003.

Wyatt, Justin. *High Concept: Movies and Marketing in Hollywood*. Austin, TX: University of Texas Press, 1994.

Yeates, Helen. "Ageing Masculinity in *NYPD Blue*: Spectacle of Incontinence, Impotence, and Morality." *Canadian Review of American Studies* 31.2 (2001): 47–56.

Young, Alison. *Imagining Crime: Textual Outlaws and Criminal Conversations*. London: Sage Publications, 1996.

Young, Iris Marion. *Justice and the Politics of Difference*. Princeton, NJ: Princeton University Press, 1990.

Ziegler, Robert. "Communication Spaces in *Hill Street Blues*." *Clues* 8.2 (1987): 79–87.

Zoglin, Richard. "Cool Cops, Hot Show." *Time* 16 Sept. 1985: 60–62.

Zoglin, Richard, & Patrick Cole. "Bochco under Fire." *Time* 27 Sept. 1993: 81.

Zynda, Thomas. "The Metaphoric Vision of *Hill Street Blues*." *Journal of Popular Film and Television* 14.3 (1986): 100–13.

INDEX

Note: 'N' after a page number indicates a note.